Climate Change Perception and Changing Agents in Africa & South Asia

Edited by

Suiven John Paul Tume
Green Care Association, Cameroon

&

Vincent Itai Tanyanyiwa
Zimbabwe Open University, Zimbabwe

Series on Climate Change and Society

Vernon Press

www.vernonpress.com

In the Americas:
Vernon Press
1000 N West Street,
Suite 1200, Wilmington,
Delaware 19801
United States

In the rest of the world:
Vernon Press
C/Sancti Espiritu 17,
Malaga, 29006
Spain

Series on Climate Change and Society

Library of Congress Control Number: 2018940778

ISBN: 978-1-62273-308-8

Table of Contents

Figures, Tables and Boxes

Figures

Tables

Boxes

Contributing Authors

Abraham Matamanda (Zimbabwe) holds a Master of Social Ecology from the Centre for Applied Social Sciences and a BSc Honours' degree holder in Rural and Urban Planning. Both degrees were obtained from the University of Zimbabwe. He is a registered full-time PhD student in the Department of Urban and Regional Planning at the University of Free State in South Africa. Currently, Abraham's PhD focuses on emerging human settlement forms and urban dilemmas nexus such as water, sanitation, public transport and security issues.

Anup K C (Nepal) received his Master's Degree in Environmental Science in 2012 and in Economics in 2014 from Tribhuvan University. He currently works as Assistant Professor of Environmental Science at Amrit Campus, Tribhuvan University, Nepal. Since 2005, he has been involved in the teaching sector in various institutions. Since 2009, he has also been involved in different development sector projects (water and sanitation, education, hydropower, transmission line, solar energy). He has conducted research on cost-benefit analysis, community forest management, carbon sequestration, ecotourism, hydropower development and climate change in different regions of Nepal.

Innocent Chirisa (Zimbabwe) is a full Professor teaching Urban and Environmental Planning courses in the Department of Rural & Urban Planning at the University of Zimbabwe. Currently, he is serving as the Deputy Dean of the Faculty of Social Studies, University of Zimbabwe. He is also serving as the Acting Chairman of the Department of Rural & Urban Planning at the University of Zimbabwe. Professor Chirisa was recently appointed as Research Fellow in the Department of Urban and Regional Planning at the University of Free State, Bloemfontein, South Africa for the period 2016 to 2020.

John Mutambwa (Zimbabwe) is a Lecturer in the Department of Linguistics at the University of Zimbabwe. His research interests include development communication, academic writing, professional communication skills, climate change communication and English language teaching methodologies. He holds a Master's of Education in Language Education, with specialisation in English, from the University of Zimbabwe.

Lun Yin (China) received his MA in Social Anthropology Studies from the Ecole Des Hautes Etudes en Science Sociale in Paris and his PhD from Minzu University of China in Beijing. He is also the Deputy Director of Centre for Biodiversity and Indigenous Knowledge (CBIK). His major field of interest is Indigenous Knowledge among mountain ethnic groups in Yunnan, which links watershed management, environment, and biodiversity. His work is dedicated to the adaptation and resilience of climate change in support of sustainable livelihoods of the mountain indigenous peoples in Yunnan province and Eastern Himalaya Mountain, China.

Mbuya Dora Nyukighan (Cameroon) holds a Bachelor's Degree in English Language from the University of Buea (1996), a Post Graduate Diploma in Education from the National Teachers' Institute Kaduna, Nigeria (2009) and a Master's degree in African Literature from the Ahmadou Bello University in Zaria, Nigeria (2012). She is a PhD research fellow on 'Gender Representations in Cameroon Drama of English Expression' at the University of Yaounde I. She is a Lecturer in the Department of English and Literary Studies, Federal University Lokoja, Nigeria.

Misiani M. Zachary (Kenya) is a Meteorologist, GIS and Remote Sensing expertise at Kenya Meteorological Department. He holds a BSc degree in Meteorology from the University of Nairobi and is an MSc GIS and RS candidate at Jomo Kenyatta University of Agriculture and Technology in the Department of Geomatic Engineering and Geospatial Information Systems (GEGIS). Before starting another Master of Meteorology at Nanjing University of Information Science & Technology NUIST, school of Atmospheric Sciences, Misiani worked previously in the Regional Center for Mapping of Resources for Development (RCMRD) in Monitoring of Environment for Security in Africa-(MESA) and Data Management and Dissemination Sections and the National Meteorological Centre office (KMD Headquarters).

Moye Eric Kongnso (Cameroon) holds a Bachelor's Degree in Geography and Environmental Management (2007) and a Master's Degree in Geography (Applied Agro-climatology, 2011), both from the Dschang. He is currently a PhD fellow researching on climate change and livelihoods in the Ndop Plains of the Bamenda Highlands of Cameroon. He is a member of a number of professional and research groups in environmental issues.

Njodzeka Gilbert Njodzeka (Cameroon) has a Diploma in Tree Nursery and Soil Conservation (1997) from Rural Training Centre (RTC), Mfonta. He is the Coordinating Manager and the Head of Project Development at Green Care Association, Shisong–Cameroon. He is a crusader in combating environmental degradation through tree planting, apiculture, water catchment protection, developing renewable energy and animal husbandry through non-conventional livestock. He has won several international awards such as Bui Family Union, New Jersey Chapter–Appreciation of Service, 2010; United Kingdom-Commonwealth Mid-Career Fellowship (2010) and United Kingdom-World Environment Radio-First Prize, Best Feedback (2008).

Nyuykongadzem Emma Dindze (Cameroon) holds a Bachelor's Degree in Geography from the University of Buea and a Master's Degree in Geography and Planning from the University of Bamenda. Her research over the years has been on the vulnerability of agro-hydrological systems to climate change in the Bamenda Highlands of Cameroon. She is also engaged in community capacity building for self-reliance in potable water supplies and health care delivery in rural areas. She is currently undertaking a professional teachers' training course in Geography at the University of Bamenda.

Rejoice Madobi (Zimbabwe) is an environmentalist with broad research interests in water resources management, poverty, and gender issues as well as climate change. Rejoice has been involved in university education for six years.

Suiven John Paul Tume (Cameroon) has a Bachelor's Degree in Geography, a Postgraduate Diploma in Education and Master's Degree in Geography all from the University of Buea. He is currently a PhD research fellow working on indigenous adaptations to climate change in the agro-hydrological systems of the Bamenda Highlands of Cameroon. He volunteers at Green Care Association, Shisong–Cameroon as a researcher on indigenous climate change adaptation. He is also a consultant and volunteer researcher at Green Future Consulting where he works on climate change vulnerability and adaptation. He is a part-time instructor in the Departments of Geography and Environmental Studies at the Catholic University of Cameroon (CATUC), Bamenda.

Susy Wandera (Kenya) holds a Bachelor's Degree in Economics from the University of California, Irvine and a Master's Degree in International Studies from the Claremont Graduate University, California. She is the Secretary of the Board and the Climate Change focal point at the Sustainable Environmental Development Watch (SUSWATCH), Kenya. She is also the Convener of the Road to COP (Conference of Parties) group in the Kenya Climate Change Working Group (KCCWG). She was a member of the *We Have Faith* COP17 campaign, where she organised the committee, provided technical support and travelled with truck 6 from Nairobi to Durban.

Takudzwa Leonard Mathende (Zimbabwe) holds a Bachelor of Social Work Honours Degree from the University of Zimbabwe and an MA in Social Impact Assessment from the University of Johannesburg. He is currently working as a Social Worker in a UK local authority. Previously, he worked in the Child Protection sector both in Zimbabwe and South Africa. His interests include social development policy and social impact assessment.

Tatenda Goodman Nhapi (Zimbabwe) possesses frontline social work experience. He has an interest and experience in social development having practised development work both for state and non-state actors. His research interests are in applied social research and social policy. At present, he is practising social work in a UK local authority. In 2015 he graduated with an Erasmus Mundus Masters in Advanced Development Social Work.

Vincent Itai Tanyanyiwa (Zimbabwe) is an Environmental Geographer/ Sustainability Practitioner with broad research interests in ecosystem services, climate change, water issues, rural-social differentiation and urban sustainability. At the most fundamental level, he is interested in understanding people's interaction with their natural environment. This broad intellectual project is undertaken, conceptually and empirically, by examining how humans live in their natural environment with non-human species in a world that is highly human-dominated. Vincent has been involved in university education over the last 12 years.

Acronyms

°C	Degree Celsius
ACT!	Act Change Transform
ADB	Asian Development Bank
ADF	African Development Fund
AfriYOCC	African Youth Conference on Climate Change
AGDP	Agricultural Gross Domestic Product
AGN	African Group of Negotiators
AN	Above Normal
AR5	Fifth Assessment Report of the Intergovernmental Panel on Climate Change
ASALs	Arid and Semi-Arid Lands
ASDEP	Association for Sustainable Development and Environmental Protection
ASEC	Association for Environmental Education-Cameroon
AUC	African Union Commission
AWGGCC	African Working Group on Gender and Climate Change
AYICC	African Youth Initiative on Climate Change
BN	Below Normal
CAEPP	Community Agriculture and Environmental Protection Project
CAFOD	Catholic Agency for Overseas Development
CAHOSCC	Committee of African Heads of State and Government on Climate Change
CAMGEW	Cameroon Gender and Environmental Watch
CAMTRACC	Cameroon Traditional Rulers Against Climate Change

CAP	Community Adaptation Planning
CARE	Cooperative for Assistance and Relief Everywhere
CAs	Communal Areas
CBOs	Community Based Organisations
CBS	Central Bureau of Statistics
CC	Climate Change
CCA	Climate Chang Adaptation
CCD	Climate Change Directorate
CCRN	Climate Change Radio Network
CEC	Committee of Executive Committee Members
CEPCSI	Centre for Environmental Protection and Community Support Initiatives
CFS	Climate Field Study
CIG	Common Initiative Group
CIRAN	Centre for International Research and Advisory Networks
CIRMAD	Centre for Indigenous Resources and Development
CoG	Council of Governors
COMINSUD	Community Initiative for Sustainable Development
COP	Conference of Parties to the United Nation Framework Convention on Climate Change
CORDAID	Catholic Organisation for Relief and Development Aid
CRIC	Committee for the Review of the Implementation of the Convention
CRM	*Changieni Rasilimali*
CRTV	Cameroon Radio and Television
CSOs	Civil Society Organisations
CYPCC	Committee of African Heads of State and Government on Climate Change's Youth Programme on Climate Change
DDC	District Development Committee

DFID	Department for International Development
DHM	Department of Hydrology and Meteorology
DRR	Disaster Risk Reduction
ED	Earth Day
EEA	European Environmental Agency
EIA	Environmental Impact Assessment
EJN	Earth Journalism Network
EMA	Environmental Management Agency
ENSO	El Nino/Southern Oscillation
ESPACE	European Spatial Planning: Adapting to Climate Events
EU	European Union
FCZ	Forest Company of Zimbabwe
FECOFUN	Federation of Community Forestry Users Nepal
FGDs	Focus Group Discussions
GCA	Green Care Association, Shisong-Cameroon
GCMs	Global Climatic Models
GEF	Global Environmental Facility
GHGs	Greenhouse Gases
GoK	Government of Kenya
GoZ	Government of Zimbabwe
HBS	Heinrich Böll Stiftung
HS & G	Head of State and Government
I/NGOs	International Non-governmental Organizations
IASSW	International Association of Schools of Social Work
ICCA	Institute for Climate Change and Adaptation
ICSW	International Council of Social Welfare

ICT	Information Communication Technology
IDBD	International Day for Biological Diversity
IDNDR	International Day for Natural Disaster Reduction
IDPEEAC	International Day for Prevention of the exploitation of the Environment in Armed Conflicts
IDPOL	International Day for the preservation of the Ozone Layer
IDRC	International Development Research Centre
IES	Institute of Environmental Science
IFSW	International Federation of Social Workers
IGBP	International Geosphere-Biosphere Programme
IIED	International Institute for Environment and Development
IK	Indigenous Knowledge
IKS	Indigenous Knowledge Systems
ILO	International Labour Organization
IMD	International Mountains Day
INDCs	Intended Nationally Determined Contributions
IPCC	Intergovernmental Panel on Climate Change
ITCZ	Inter-Tropical Convergence Zone
ITK	Indigenous Traditional Knowledge
KAPSLMP	Kenya Agricultural Productivity and Sustainable Land Management Project
KCCWG	Kenya Climate Change Working Group
KCF	Kenya Climate Change Forum
KICD	Kenya Institute for Curriculum Development
KIIs	Key Informant Interviews
KYCN	Kenya Youth Climate Network
LAPA	Local Adaptation Plan for Action

LDCs	Least Developed Countries
LDN	Land Degradation Neutrality
LEK	Local Ecological Knowledge/ Local Environmental Knowledge
LK	Local Knowledge
LTK	Local Traditional Knowledge
LWP	Lima Work Programme
MECCOD	Media Synergy for the Promotion of Biodiversity Conservation and Community Development against Climate Change
MENR	Ministry of Environment and Natural Resources
MEWC	Ministry of Environment, Water and Climate
MIFACIG	Mixed Farming Common Initiative Group
MOPE	Ministry of Population and Environment
MSD	Meteorological Service Department
MSDZ	Meteorological Services Department, Zimbabwe
NAP	National Adaptation Plan
NAPA	National Adaptation Programmes of Action
NCA	Norwegian Church Aid
NCCAP	National Climate Change Action Plan
NCCCK	National Climate Change Consortium of Kenya
NCCRS	National Climate Change Response Strategy
NCCSP	Nepal Climate Change Support Programme
NDC	National Determined Contribution
NEST-Cam	Network of Environmental Stakeholders Cameroon
NGEC	National Gender and Equality Commission
NGO	Non-Governmental Organisation
NIMBY	Not In My Back Yard

NSAs	Non-State Actors
NTNC	National Trust for Nature Conservation
NYCCC	National Youth Conference on Climate Change
OECD	Organisation for Economic Co-operation and Development
PACJA	Pan-African Climate Justice Alliance
PANERECC	Parliamentary Network on Renewable Energy and Climate Change
PAPNCC	Pan-African Parliamentarians Network on Climate Change
PPCR	Pilot Programme for Climate Resilience
RD	Rural Development
RDC	Rural Development Councils
RDPs	Rural Development Programmes
REDD	Reduced Emissions for Deforestation and Forest Degradation
RuWCED	Rural Women Centre for Education and Development
RYCCC	Regional Youth Conference Climate Change
SADC	Southern African Development Community
SARCOF	Southern African Regional Climate Outlook Forum
SBI	Subsidiary Body of Implementation
SCFs	Seasonal Climate Forecasts
SDG	Sustainable Development Goal
SHUMAS	Strategic Humanitarian Services
SIRDEP	Society for Initiatives in Rural Development and Environmental Protection
SPCR	Strategic Program for Climate Resilience
SURUDEV	Sustainable Run for Development
SUSWATCH	Sustainable Environmental Development Watch
SYFA	Save Your Future Association

TCPL	Total Consumption Poverty Line
TEK	Traditional Ecological Knowledge/Traditional Environmental Knowledge
TIMB	Tobacco Industry Marketing Board
TK	Traditional Knowledge
UN	United Nations
UNCCD	United Nations Convention to Combat Desertification
UNCT	United Nations Country Team
UNDP	United Nations Development Programme
UNEP	United Nations Environment Programme
UNESCO	United Nations Educational, Scientific and Cultural Organization
UNFCCC	United Nations Framework Convention on Climate Change
UNICEF	United Nations Children's Fund
UZ	University of Zimbabwe
VDC	Village Development Committee
WB	World Bank
WCD	World Cities Day
WDCDD	World Day to Combat Desertification and Droughts
WED	World Environment Day
WFD	World Food Day
WHD	World Habitat Day
WMD	World Meteorology Day
WMO	World Meteorological Organization
WOD	World Oceans Day
WTD	World Tourism Day
WWD	World Water Day

WWF	World Wide Fund for Nature
ZERO	Zimbabwe Regional Environment Organisation
ZIMVAC	Zimbabwe Vulnerability Assessment Committee

Definitions

Africa is the world's second largest and second most populous continent with a land area about 30.3 million km^2 with 1.2 billion people. Africa is surrounded by the Mediterranean Sea to the north, both the Suez Canal and the Red Sea along the Sinai Peninsula to the northeast, the Indian Ocean to the southeast and the Atlantic Ocean to the west.

Agro-pastoralism is a form of social organization based on the growing of crops and the raising of livestock as the primary means of economic activity.

Cameroon is a country in Central Africa, bordered by Nigeria to the west; Chad to the northeast; the Central African Republic to the east; and Equatorial Guinea, Gabon and the Republic of the Congo to the south. The capital of Cameroon is Yaoundé.

Climate change refers to any change in climate over time, whether due to natural variability or as a result of human activity (IPCC,2007).

Climate Field School (CFS) a group formed by local experts and normal villagers with traditional knowledge in Kenya

A **climate model** is a numerical representation of the climate system based on the physical, chemical, and biological properties of its components, their interactions and feedback processes, and accounting for all or some of its known properties.

Climate variability refers to variations in the mean state and other statistics (such as standard deviations, the occurrence of extremes, etc.) of the climate on all temporal and spatial scales beyond that of individual weather events. Variability may be due to natural internal processes within the climate system (internal variability), or to variations in natural or anthropogenic external forcing (external variability) [IPCC, 2007].

El Nino is the widespread warming of the upper ocean in tropical Pacific. This results in increasing rainfall in the eastern Pacific and decreasing rainfall over the western Pacific and Southern Africa.

El Nino Southern Oscillation (ENSO) denotes a coupled atmospheric system that links change in atmospheric pressure and sea temperature over the southern Pacific Ocean

Environmental change is the disturbance of the environment mainly caused by anthropogenic influences and natural ecological processes resulting in disasters etc.

Greenhouse gases are those gaseous constituents of the atmosphere, both natural and anthropogenic, that absorb and emit radiation at specific wavelengths within the spectrum of infrared radiation emitted by the Earth's surface, the atmosphere, and clouds.

Indigenous knowledge is a cumulative and complex body of knowledge, practices, and representations that are maintained and developed by peoples with extended histories of interactions with the natural environment (Hiwasaki *et al.*, 2014)

Intergovernmental Panel on Climate Change (IPCC) is a technical and intergovernmental body under the auspices of the United Nations, set up and dedicated to the task of providing the world with an objective, scientific view of climate change and its socio-economic and political impacts.

Kenya is a country in East Africa bordered by Tanzania to the south and southwest, Uganda to the west, South Sudan to the north-west, Ethiopia to the north and Somalia to the north-east. Kenya covers 581,309 km^2

Livelihood comprises the capabilities, assets (including both material and social resources) and activities required for a means of living (Chambers & Conway, 1991).

Mass communication is the transmission of information and ideas to a largely dispersed, heterogeneous audience simultaneously (Okenwa, 2002)

National Climate Change Action Plan (NCCAP) is a body in Cameroon aimed at building the capacity of socio-economic actors to adjust to climate change (UNDP, 2009)

Nepal whose capital is Kathmandu is a landlocked central Himalayan country in South Asia. It borders with China in the north and India in the south, east, and west. Nepal is the largest sovereign Himalayan state.

Not in My Back Yard (NIMBY) is a colloquialism signifying one's opposition to the locating of something considered undesirable in one's neighborhood.

Seasonal climate forecasts (SCF) are produced operationally in tercile-probabilities of the most likely categories, e.g., below-, near- and above-normal rainfall. Inherently, these are difficult to translate into information useful for decision support in farming.

South Asia is the southern region of the Asian continent and is made up of countries lying between the Himalaya range of mountains and the Indian Ocean (north to south) and between the Ganga and Indus river valleys (east to west).

The **Conference of Parties**, known as COP, is the decision-making body responsible for monitoring and reviewing the implementation of the United Nations Framework Convention on Climate Change made up of the 197 nations and territories – called Parties – that have signed on to the Framework Convention. The COP has met annually since 1995.

The **Eastern Himalayan region** refers to the area lying between 82.70°E and 100.31°E longitude and 21.95°N to 29.45°N latitude, covering a total area of 524,190 sq.km. The region extends from the Kaligandaki Valley in central Nepal to northwest Yunnan in China, also encompassing Bhutan, the northeastern states and north Bengal hills in India, southeastern Tibet, and parts of northern Myanmar.

The **Inter-Tropical Convergence Zone (ITCZ)** is an expanse of air where the northeast and southeast trade winds converge and characterized by much convection activity resulting in rainfall during summer months in Southern Africa.

Tibet is a region covering a greater of the *Tibetan* Plateau in Central Asia. It is the homeland of the *Tibetan* people and other ethnic groups such as Monpa, Qiang, and Lhoba peoples. It is also occupied by Han Chinese and Hui people.

Vulnerability refers to the diminished capacity of an individual or group to anticipate, cope with, resist and recover from the impact of a natural or man-made hazard (UNISDR, 2007)

Zanba, a staple food of Tibetan people, is *Zanba*, a kind of dough made with roasted highland barley

Zimbabwe is a landlocked country located in southern Africa, between the Zambezi and Limpopo Rivers, bordered by South Africa, Botswana, Zambia, and Mozambique. The capital and largest city is Harare.

Chapter One
INTRODUCTION

Suiven John Paul Tume, *Green Care Association, Cameroon*

Vincent Itai Tanyanyiwa, *Zimbabwe Open University, Zimbabwe*

Climate change is a socio-ecological system which cannot be understood and evaluated by relying on physical sciences. This is mainly because its effects trickle down to the lowest level of society, particularly in rural communities of developing countries. This stems from the fact that rural communities rely on climate-sensitive sectors that include agriculture and other primary activities for their livelihoods. Agriculture in most rural areas of Sub-Saharan Africa and South Asia is rain-fed. Thus, the peoples' perceptions about the changing climate and environmental conditions form the basis for their decision-making with regards the agricultural calendar and seasonal changes in water resources. Climate variability and change at the local level gives rise to the attainment of insights into the experiences of local people that are hardly reflected in hard-core scientific studies and models. The significance of indigenous climate perceptions is critical to understanding the socio-cultural context within which environmental change occurs. Local climatic changes are a prelude to global climatic processes. Many stakeholders complement indigenous perceptions to make informed choices in order to improve on livelihoods of the affected communities. Amongst these stakeholders in the climate-environmental communication chain are local media outlets such as community radios, newspapers, Common Initiative Groups, Non-governmental Organisations, community-based organisations, socio-cultural groups, municipalities and or councils and government departments.

The situation of climate change is clear and urgent. This is evidenced by the increasing heat-trapping gasses in the atmosphere and the associated heat waves, forest fires, and heavy downpours, sea level rise among others. There is a need for effective action especially in the way in which climate science is disseminated in order to curtail the profound and irreversible consequences, especially in developing countries. The actions we take (or do not take) now will impact all life on earth for thousands of years to come. If we fail to act quickly, we risk leaving our children a problem they cannot solve. This especially important if we take into consideration the adage;

We Do Not Inherit the Earth from Our Ancestors;
We Borrow It from Our Children

We have to undertake to cherish and not damage the environment in which we live, not because we are duty-bound by the dictates of humanity but because we care and love those who will live beyond us. Climate change is already underway hence the need for current and future generations to adapt to some extent. Obviously, this depends on the level of imminent warming. A business-as-usual approach will lead to self-destruction and will result in perilous climate change and potentially calamitous impacts. There is a need for effective communication, as well as policies and technologies to adapt to climate change. It is high time we get started to disseminate climate change information in a way that is easily understood by common people.

Humanity is having a profound impact on climate change. There is an urgent need to bring about significant and lasting changes in our economics and societies. Understanding how people and societies develop awareness is essential for the design of effective communication strategies on this issue. In order to tackle the climate change challenge on time, communication research is exploring the core of public awareness engagement. The once simplistic models of communication construed as a one-way process (from messenger to a passive audience) are evolving. Decades of environmental activism and communication have contributed to a more sophisticated understanding of communication processes which takes into account underlying factors at the basis of individual and or communal decisions and actions, such as experiences, mental or cultural models, and relational dynamics.

Global Climate Change

Climate change analysis assesses the likelihood that a particular extreme weather event has been made more or less likely as a result of anthropogenic causes (Budimir and Brown, 2017). The Earth's climate is driven by interactions between the atmosphere, oceans, lithosphere, terrestrial and marine biospheres (Chakraborty *et al.*, 2000; International Geosphere-Biosphere Programme-IGBP, 2008). The increasing build-up of greenhouse gases (GHG) in the atmosphere and escalating concentrations of natural and anthropogenic aerosols have detectable effects on the climate system (Sivakumar *et al.*, 2005). Changes in atmospheric composition due to increasing concentrations of greenhouse gases (mainly carbon dioxide, methane, nitrous oxide), changes in land cover and unsustainable agricultural practices are responsible for warming the earth's surface (Intergovernmental Panel on Climate Change-IPCC, 2007; Collier *et al.*, 2008; Yanda and Mubaya, 2011; Omambia et *al.*, 2010). Although there are debates among scholars with regard to whether climate change is induced by anthropogenic activities or is as a result of natural climate variability, the balance of scientific opinion is that changes in the composition of the atmosphere are mainly attributed to anthropogenic activities (IPCC, 2001; 2007; 2014).

The total anthropogenic GHG emissions have continued to increase from 1970 to 2010, with the highest amount noted between 2000 and 2010 (IPCC, 2014). This report further notes that the release of CO_2 into the atmosphere from the burning of fossil fuels and industrial activities contributed about 78% of the total GHG emissions from 1970 to 2010, with a similar increase from the period 2000 to 2010 (IPCC, 2014). Rising temperatures heat the land mass and the surrounding oceans, causing increases in surface temperatures and changes in precipitation, which are important drivers of global climate change (Collier *et al.*, 2008; Challinor *et al.*, 2007; Boucher, 1999). In spite of the fact that trends and patterns of climate change projections are generally consistent, they are subject to varying degrees of uncertainty due to limitations in measurements and knowledge of the interactions between earth-atmosphere systems (Adger *et al.*, 2003; Challinor *et al.*, 2007).

Global temperatures near the earth's surface increased by 0.74°C from 1906 to 2005 and are estimated to increase by about 6.4°C on average during the 21st century (IPCC, 2007; Boucher, 1999). Recent evidence and predictions from computer models indicate that climate change is accelerating and will lead to wide-ranging shifts in climate variables (IPCC, 2007; Chaudhary and Aryal, 2009). The global climate models (GCMs) project an increase in the global mean temperature of between 1.5 and 5.8°C by the end of 2100, which is attributed to population growth, energy use and land cover changes (Elum *et al.*, 2017). The IPCC (2014) argues that the previous three decades, from 1983 to 2012, are most likely to be the warmest periods of the last 1,400 years in the Northern Hemisphere, whereas the global average surface temperature data for the land and sea combined show a warming of 0.85 [0.65-1.6] °C, over the period from 1880 to 2012.

Increased temperature affects ecosystems and biological communities (Chaudhary and Aryal, 2009; IPCC, 2014; United Nations Environment Programme-UNEP, 2010; Seddon *et al.*, 2016). Some of the effects include droughts, floods, frequent fires, species shifts and sensitive diseases increase (IPCC, 2001; 2007; 2014). Researchers suggest that with the warming conditions, precipitation patterns are likely to change, with increases up to 20% projected in some parts of the world, although drought conditions will also be exacerbated, particularly in Africa (Rosenzweig *et al.*, 2001; Collier *et al.*, 2008; Hulme *et al.*, 2001; Collins *et al.*, 2013; Toulmin *et al.*, 2005; Toulmin and Huq, 2006). Changes in temperature and precipitation are also projected to influence extreme weather events (floods, drought), food production and prices;

water availability and access (FAO, 2014; IPCC, 2007; Omambia *et al.*, 2010; Mathews *et al.*, 2016; Collins *et al.*, 2013).

Socio-economic impacts of climate variability are noteworthy and will impact humans through a variety of direct and indirect ways (Heltberg *et al.*, 2009; IPCC, 2007; Elum *et al.*, 2017). Generally, the impacts of climate variability and change are projected to have enormous and devastating global consequences, but the most adverse impacts are predicted to occur in developing countries due to their fewer resources and technological challenges to cope with and adapt to the changing conditions (Elum *et al.*, 2017; Omambia *et al.*, 2010). This is due to their location at fragile, marginal and vulnerable environments and over-dependence on agriculture, which is a climate-sensitive sector (Stern, 2007; IPCC, 2007; Omambia *et al.*, 2010; Oppenheimer, 2014). Vulnerability to climate variability and change, in turn, poses multiple stresses to economic growth and poverty reduction strategies in Africa (IPCC, 2007; Stern, 2011; Elum *et al.*, 2017). This calls for vigorous adaptation strategies.

Climate change adaptation (CCA) is managing the inevitable (United Nations Framework Convention on Climate Change[UNFCCC] (UNFCCC, 2006, 2011; IPCC, 2007, 2014). It is the adjustments of ecological and socio-economic systems in response to actual or expected climatic stimuli and their effects (Peloquin and Berkes, 2009; IPCC, 2007). Climate adaptation is the change in processes, practices, and structures to moderate potential damages or to benefit from opportunities associated with climate change (Smit and Pilifosova, 2001; IPCC, 2007, 2014). Adaptation activities span through observation of climate change impacts, assessment of impacts and vulnerability, planning for adaptation, implementation of adaptation plans, monitoring, and evaluation of adaptation actions (UNFCCC, 2011; Oppenheimer, 2014). Climate adaptation takes the following directions: anticipatory (proactive), planned, reactive, autonomous (spontaneous), private and public (IPCC, 2007, 2014; Snow and Prasad, 2011; Töpfer and Hunter, 2002; Rhodes *et al.*, 2014; OECD, 2009). Individuals and nature itself often adapt to climate change impacts without any external help (Töpfer and Hunter, 2002). In many cases, people need to plan how to minimise the costs of negative impacts and maximise the benefits from positive impacts. Planned adaptation can be launched prior to, during or after the onset of the actual consequences of climate change (IPCC, 2007; UNFCCC, 2006).

Understanding of the magnitude of the adaptation challenge at a local or global scale is often constrained by a limited understanding of how adaptation is taking place (Eisenack and Stecker, 2013). This is because many studies report on vulnerability assessments of natural systems and not adaptation actions (Philander, 2008). Climate change is rarely the sole motivator for adaptation action because extreme events such as floods and droughts are important adaptation stimuli across sectors (Smit and Pilifosova, 2001; Kolawole *et al.*, 2014; FAO, 2013; Mathews *et al.*, 2016). Pro-active adaptation is the most commonly reported adaptive response (IPCC, 2007). Adaptation action is more frequently reported in developed nations unlike in middle and low-income countries mainly due to poor climate change communication strategies. There also is limited reporting on adaptations for vulnerable people such as women, the elderly or children (Berrang-Ford *et al.*, 2010).

Stakeholders in Climate Change Communication

Communication of extreme weather events in the immediate aftermath of an event provides a window of opportunity to inform, educate and affect a change in attitude or behaviour in order to cope, adapt and or mitigate or prepare for climate change. Prompt access to information can help decision makers to ensure that appropriate adaptation and investment decisions are taken. Effective communication of climate change information is critical to ensuring that decision-makers at all levels do indeed understand and are able to act upon such information (Budimir and Brown, 2017).

Climate change communication has lately become conspicuous in society from movies to grassroots movements (Nerlich *et al.*, 2009; Ward and Menezes, 2008). It is a global problem with widespread impacts. It is crucial that climate change messages are communicated effectively by different stakeholders including the media, Non-governmental Organisations(NGOs), socio-cultural groups, municipal councils and government institutions (European Spatial Planning: Adapting to Climate Events-ESPACE, 2007). Climate change communication and its impact on the general public have proliferated in communication and related disciplines since the 1990s (Nerlich *et al.*, 2009). In spite of this, society remains vulnerable to climate variability and change. This raises questions about the effectiveness of communication efforts and the ability of audiences to implement adaptation in response to these communications (Moser, 2010; Monroe *et al.*, 2015). This concerns persuasiveness of the messages, the structure of society and considerations of the extent to which the public is receptive to make an effective change (Nerlich *et al.*, 2009; IPCCC, 2014; UNFCCC, 2006). The combination of natural and human-induced factors accelerate climate change, and consequences abound [interruptions in agriculture, surges in the frequency of droughts and floods] (Filho, 2009). Since human-induced climate change first emerged on the public agenda in the mid-1980s, public communication of climate change and more recently the question of how to communicate it most efficiently have witnessed a sharp rise (Moser, 2010; Monroe *et al.*, 2015; Nerlich *et al.*, 2009).

The Fifth Assessment Report (AR5) of the IPCC is widely recognised by academics, journalists, communications experts, governments and civil society organizations to have marked strides in the approach the IPCC communicates its agenda (IPCC, 2016). The communication of AR5 saw both greater professionalism at all stages of the process and greater breadth and diversity in the subsequent outreach accomplishments. The results of this can best be seen in the Paris Agreement of the United Nations Framework Convention on Climate Change (UNFCCC) reached at the Conference of Parties in December 2015 (COP-21). This agreement is based on assessments that the IPCC communicated to negotiators through the Structured Expert Dialogue and UNFCCC meetings. The IPCC has also experienced growing calls from policymakers and other users to do more with its communications. Some improvements to IPCC communications came not from the communications team but from the authors of AR5-the use of headlines statements in the Working Groups I, II & III contribution to AR5 and the Synthesis Reports. Other improvements for AR5 included:

- Responding to media questions before completion of the reports; Media workshops to clarify the workings of the IPCC and how it produces assessments; Making IPCC communications more professional by working with external communications specialists; Media training for bureau members and authors; Systematic planning of interviews with a range of authors, both face-to-face and remotely; Arranging facilities for broadcasters; Production of scientifically rigorous but compelling videos overseen by the working group co-chairs and IPCC Chair; Ambitious programme of outreach activities all over the world; Cooperation with third parties producing versions of the report targeting specific sectors in specific regions; Use of social media to publicise IPCC findings and outreach activities.

This gives the IPCC a strong foundation to build on for its future climate change communications.

Climate Change Agents

The poor must be able to sustainably manage and benefit from the rich and abundant natural resources to address poverty and promote inclusive growth (Manjengwa, 2012). Moreover, the Human Factor approach usage is relevant for discourse analysis in this book. The main claim of the Human Factor approach as propounded by Mararike (2014) is that no nation or country can sustain its development activities without people who are patriotic, reliable, committed and disciplined. For this social workers are competently fit for this purpose as change agents with core competencies to make tailor-made rural development programmes. Indigenous Knowledge Systems (IKS) need to be harnessed comprehensively to facilitate social workers led pro-poor rural development interventions mitigating climate change. IKS consists of a body of knowledge, beliefs, traditions, practices, and institutions developed and sustained by indigenous, peasant and local communities in interaction with their biophysical environment (Bhatasara and Mandizadza, 2014). Though diagnostic than prognostic, indigenous indicators, such as trees physiological behaviour, behaviour of frogs and sounds of birds, are related to the response of certain animals and plants to the already prevailing weather form indigenous knowledge for climate forecasting. Whilst IKS has been a source of resilience; climate change provokes serious threats to both livelihoods and the nature of indigenous climate knowledge inherent within a given locality (Bhatasara and Mandizadza, 2014).

Indigenous Knowledge Support for Decision-Making

Indigenous knowledge is a cumulative and complex body of knowledge, practices and representations that are maintained and developed by peoples with extended histories of interactions with the natural environment (Hiwasaki *et al.,* 2014; Macchi *et al.*, 2008; Egan, 2013; Nakashima *et al.*, 2012; Kolawole *et al.,* 2014; Rhodes *et al.,* 2014., Tanyanyiwa & Chikwanha, 2011). These systems are part of a complex that includes language, attachment to place, spirituality and perception of worldview (Macchi *et al.*, 2008). Notwithstanding a variety of terminology used to refer to local environmental knowledge (LEK), such as traditional local knowledge(TLK), aboriginal knowledge(AK), folk knowledge(FK), traditional wisdom(TW), traditional science(TS), people's science(PS), traditional ecological knowledge (TEK), indigenous knowledge (IK), local knowledge(LK), rural peoples knowledge(RPK), ethno-biology/ethno-botany/ethno-zoology, ethno-science; folk science(FS), traditional knowledge(TK); indigenous traditional knowledge (ITK), indigenous technical knowledge(ITK), traditional environmental knowledge (TEK), rural knowledge(RK), indigenous science(IS) (Hiwasaki *et al.*, 2014; Kolawole *et al.*, 2014, 2016; Fernández-Llamazares *et al.*, 2015; Mwaura, 2008; Hiwasaki *et al.*, 2014; Nakashima *et al.*, 2012; Krishna, 2011; Briggs, 2005; Thompson and Scoones, 1994; Senanayake, 2006; Beckford and Barker, 2007; Scoones, 1998; Ellen and Harris, 1996; Odero, 2011; Nakashima *et al.*, 2012), all the terminology has similar meanings and is used interchangeably to refer to the local environmental or traditional knowledge and skills held by indigenous people, developed outside the formal scientific domain, embedded in culture and steeped in tradition through oral tradition (Sen, 2005; Beckford and Barker, 2007; Odero, 2011).

Indigenous environmental knowledge is built up by a group of people through generations of living in close contact with nature (Centre for International Research and Advisory Networks-CIRAN, 2001; International Labour Organization-ILO, 2017). It is the knowledge used by local people to make a living in a particular environment (Kolawole *et al.*, 2014). However, indigenous knowledge is much more complex (Mwaura, 2008). Land management under indigenous knowledge involves a number of farming technologies that have repercussions across the whole spectrum of conservation (Mwaura, 2008). These include such practices as slash-and-burn, shifting cultivation, use of grass strips, intercropping, selective cultivation, and a number of other technologies and practices that seek to optimize food production under varying environmental conditions (Emgeru *et al.*, 2012). In addition, many of the communities surveyed combined cultivation with livestock rearing (Rhodes *et al.*, 2014).

The idea that LEK can provide strategies for adaptive management in the face of Global Environmental Change (GEC) Macchi, 2008; Hiwasaki *et al.*, 2014). LEK is gaining worldwide credence and recognition not only in academic circles but also at the political level (Fernández-Llamazares *et al.*, 2015). For instance, various agencies and bodies of the United Nations (UN), including the IPCC, recommend investigating local people's place-based knowledge for increasing resilience in a changing global environment (Macchi, 2008; Hiwasaki *et al.*, 2014; Wong-Parodi *et al.*, 2010; Fernández-Llamazares *et al.*, 2015). The intricate interaction between local peoples and their surrounding environments has resulted in detailed LEK that has proved to be pivotal in allowing societies to subsist in a wide range of environments and to adapt to social-ecological changes (Briggs and Moyo, 2012). LEK is contended to be *'adaptive'* because it reacts to the ever-changing nature of social and environmental conditions (Fernández-Llamazares *et al.*, 2015).

Traditional societies in many cases have built up knowledge over long periods about changes in the environment and have developed elaborate strategies to cope with climate variability and change (Macchi *et al.*, 2008; Hiwasaki *et al.*, 2014; Kolawole *et al.*, 2014; CIRAN, 2001). Traditional knowledge systems in mitigation and adaptation have for a long time been neglected in climate change policy formulation and implementation (Emgeru *et al.*, 2012; Gyampoh *et al.*, 2009; Waha *et al.*, 2013). It is only recently that indigenous knowledge has been taken up into the climate change discourse (Miller *et al.*, 2014). Traditional and indigenous peoples have survived over long periods to many kinds of environmental changes, including climate change, have valuable lessons to offer about successful and unsuccessful adaptations which could be vital in the context of climate change (Nakashima *et al.*, 2012; Rhodes *et al.*, 2014).

Indigenous and traditional peoples' representatives have been participating at UNFCCC COP since 1998 (UNFCCC, 2011). On several occasions, they have expressed their concerns and discontentedness related to the implications of climate change policies on their livelihoods and cultures (Nakashima *et al.*, 2012). Since 2001, indigenous peoples' organisations (IPOs) have been acknowledged as a constituency in climate change negotiations within the UNFCCC (Fernández-Llamazares *et al.*, 2015). However, indigenous peoples' organizations are still waiting for the approval of an ad hoc Working Group on Indigenous Peoples and Climate Change by the UNFCCC, allowing them to participate in the COP actively (Macchi, 2008; Hiwasaki *et al.*, 2014).

People living in harsh environments often have knowledge of local climatic conditions and extreme events going back generations and have developed effective strategies for adaptation, mitigation, and resilience (International Institute for Environment and Development-IIED, 2015a; Swiderska and Malmer, 2016; Mathews *et al.*, 2016). Vulnerable people's knowledge and needs should inform local and national adaptation responses, particularly the Intended Nationally Determined Contributions (INDCs) that are the UNFCCC's current focus (OECD, 2009). Not building on local knowledge brings the risk of maladaptation, or inappropriate responses (IIED, 2015a). Local and traditional knowledge must also be better-used in assessments such as the IPCC, which remains a key influence on government policymaking (IIED, 2015a, b; Fernández-Llamazares *et al.*, 2015).

Traditional African cultures comprise an institution of *'rainmakers'* who would not as much invoke the rains, but anticipate them based on *ethno-meteorology* (Kolawole *et al.*, 2014). *Ethno-meteorology* is an indigenous way of forecasting and interpreting local weather conditions in a given locality (Kolawole *et al.*, 2014; Miller *et al.*, 2014). Innovation has eroded the knowledge system of community people to the extent that these local custodians are branded as *'backward charlatans'* (Macchi *et al.*, 2008). Apart from the labelling and stigmatisation experienced by small farmers, abrupt changes in weather patterns, which also interrupts natural indicators used in ethno-meteorology, continue to overwhelm the smallholders in their present environments (Emgeru *et al.*, 2012).

Use of Weather and Climate Information for Decision-Making

The supply of climate information consists of various actors responsible for the collection of weather data, analysis, and generation of forecasts, to dissemination and use. In many countries particularly developing ones' systematic observations of the weather and climate system are usually the responsibility of national meteorological centres which are usually government departments. Meteorological Services provide information on emissions of greenhouse gases (GHGs) and information on interventions required to react to the adverse impacts of climate change. In most Asian countries also, the meteorological or hydro-meteorological department of the government is responsible for collecting, processing and supplying data as well as maintaining the infrastructure of the systematic observation system (UNFCCC, 2007). However, other institutions such as non-governmental organisations (NGOs), research centres and agricultural extension services also play a pivotal role.

Weather information is commonly given as seasonal forecasts. This refers to the total amount of rainfall expected during the season or over a given time period (Ziervogel and Calder, 2003). Inter-annual variability in Southern Africa is largely influenced by the El Nino Southern Oscillation (ENSO) phenomenon. ENSO denotes a coupled atmospheric system that links change in atmospheric pressure and sea temperature over the southern Pacific Ocean. El Nino is the widespread warming of the upper ocean in tropical Pacific. This results in increasing rainfall in the eastern Pacific and decreasing rainfall over the western Pacific and Southern Africa. In Southern Africa, most rain is brought by the Inter-Tropical Convergence Zone (ITCZ). The ITCZ is an expanse characterized by much convection activity resulting in rainfall during summer months. The likelihood of specific weather regimes is affected by sea surface temperatures, ice cover and land-sea characteristics. These boundary conditions mostly evolve slowly, and so can be used to provide a forecast of different weather regime probabilities (Mason *et al.*, 1996). Reliability and potential use of seasonal forecasts differ according to geographical region, that is, according to oceanic-atmospheric circulation, and according to the political and socio-economic context that shapes its diffusion and application. Weather forecasts are issued in base on three categories that are: below normal (BN) is when an area receives below 75% of its long-term average for the sub-season, near normal (N) is when an area receives between 75% and 125% of its long-term average in a sub-season. And above normal (AN) is when an area receives more than 125% of its long-term mean for a particular sub-season. These forecasts give probable information on the forthcoming climate of time scales of 3 to 6 months (Manatsa *et al.*, 2012).

Historical, current and future weather and climate information are necessary for climate change projections and modelling. Seasonal weather forecast and climate information (the beginning and end of each year's rainfall season have increasingly become more important as the climate scenario is increasingly becoming more and more unpredictable. This information according to World Meteorological Organization (2010) translates into:

- adequacy of rainfall amounts and deficits and excesses,
- early warnings of potentially poor seasons or disasters, for example, floods and droughts.

The information is necessary for the farmers' decision-making process that is, 'to inform key actions for general planning questions, such as when to start planting, knowing how much to diversify, knowing which crops to plant, and the likelihood of meeting or failing to meet quotas World Meteorological Organization (2010). Accurate weather forecasts and early warning increase farmer preparedness and leads to better social, economic and environmental outcomes within agricultural production systems. Important farm level decisions are based on the forecasts, and these decisions include a choice of cropping pattern, investment in fertilizers and pesticides, choice of the period for planting and plant population density. It is important for the forecasts (which are probabilistic in nature) to be well understood by the farmers if they are to be fully incorporated in farming decisions. Seasonal forecasts, therefore, give farmers an opportunity to adopt the improved technology, intensify production, replenish soil nutrients and invest in more profitable enterprises when climatic conditions are favourable and to more effectively protect their families and farms against the long-term consequences of adverse extremes. This is quite critical because crop production is predominantly rain-fed and seasonal rainfall is highly variable. Advance warning systems of extreme rainfall events should be given priority especially to the vulnerable farmers.

Climate Change Audiences

Attention paid to the audience needs and differences in climate change communication has been limited historically (Moser, 2010). Thus, the target audience has to be selected, and consideration has to be given as to how the environmental issue at hand impacts on their lives. The utility of considering the needs of the audience is clearly espoused by Rose *et al.*, (2005) who opine that effective climate change communication ought to tap into the motivations of the target audiences. While communities may be unfamiliar with climate change terms, they do know what climate change is, and this was evident through all the regions visited. Other stakeholders reached out to were faith-based, humanitarian and the wider non-state actors' constituencies, as well as private sector in all the regions the hearings, took place. Understanding end-user concerns and interests, core values, pre-existing knowledge or lack of knowledge, misconceptions, and beliefs can enable tailored communications (Budimir and Brown, 2017). For instance, in weather forecasting, useful information aims to meet recipient needs in terms of climate parameters, spatiotemporal resolution, and accuracy of the communication. Technical language must be reduced to enable users to access the most relevant information as quickly as possible. Communication approaches need to take into account how people make decisions and process information, particularly when dealing with challenging concepts such as risk, climate change, and disasters. Stakeholder values, attitudes, concerns, knowledge of climate change, language, personal and social aspirations, amongst other characteristics, should all be taken into consideration when determining what and how to communicate (Budimir and Brown, 2017).

Climate Change Perception and Changing Agents in Africa and South Asia

This book presents first-hand experiences of the concept of climate change perception in Africa and South Asia. It consists of eight core chapters drawn from Cameroon, Kenya, Zimbabwe, Nepal and East Himalayas. Chapter 1 and 10 are the introductory and conclusion chapters respectively. The Cameroonian experience is based on stakeholders in climate change communication. In Kenya, the experience of climate change communication focuses on practical field experiences based on climate change hearings. The cases from Zimbabwe are based on the role of social workers in climate change adaptation and mitigation, the use of weather information for decision-making and game changers in climate communication. The role indigenous knowledge and community-based research in climate change adaptation in East Himalayas as well as the perception of climate change in Nepal present first-hand research from South Asia. This book explores the driving factors for change on the dynamic and complex system of individuals and / or communities, their role as forces to take decisions/actions towards sustainability and their potential as tools to design climate change communication strategies. This book reflects current dynamism in the interdisciplinary study of climate change communication by contributing to the establishment of new models of communication by filling the gap between the perception of climate change as an external problem and peoples role as change agents and how they addressing the following issues; psychology of climate change communication, engagement of audiences in climate issues and solutions (messengers, messages, media...), vulnerability of communities (or resilience) to impact, the perception and understanding of the problem of climate change, current state and underlying drivers of public beliefs and attitudes about climate change, understanding of target audiences, invisible connections: from motivations to actions and finally the basis for systemic changes.

References

Adger, W.N., 2003, Social Capital, Collective Action and Adaptation to Climate Change. Economic Geography, (79) 387-404.

Beckford, C. & Barker, D., 2007, The Role and Value of Local Knowledge in Jamaican Agriculture: Adaptation and Change in Small-scale Farming. Geographical Journal, (173), 118-128.

Berrang-Ford, L., Ford, J.D. & Paterson, J., 2010, Are we Adapting to Climate Change? Global Environmental Change, 1-9.

Bhatasara, S. & Mandizadza, E. J., 2014, Climate Change and Indigenous Knowledge in Zimbabwe: A human factor approach. In: C. Mararike, ed. Land: An Empowerment Asset for Africa: The Human factor approach. Harare: University of Zimbabwe Publishers, 83-102.

Boucher, G., 1999, Global Warming. In Pacione, M. (Ed., 1999): Applied Geography: Principles and Practice, An Introduction to Useful Research in Physical, Environmental and Human Geography, Routledge, London and New York, 21-33.

Briggs, J. & Moyo, B., 2012, The Resilience of Indigenous Knowledge in Small-scale African Agriculture: Key Drivers. Scottish Geographical Journal, (128), 64-80.

Budimir, M. & Brown, S., 2017, Communicating Extreme Weather Event Attribution: Research from India and Kenya. Schumacher Centre, Warwickshire.

Centre for International Research and Advisory Networks, CIRAN, 2001, Indigenous Knowledge and Development Monitor. CIRAN, The Hague (9-3), 1-40.

Chakraborty, S., Tiedemann, A. & Teng, P., 2000, Climate Change: Potential Impact on Plant Diseases. Environmental Pollution, (108), 317-326.

Challinor, A., Wheeler, T., Garforth, C., Craufurd, P. & Kassam, A., 2007, Assessing the Vulnerability of Food Crop Systems in Africa to Climate Change. Climatic Change, 83, 381-399.

Chaudhary, P. & Aryal, K.P., 2009, Global Warming in Nepal: Challenges and Policy Imperatives. Journal of Forest and Livelihood, (8), 5-14.

Collier, P., Conway, G. & Venables, T., 2008, Climate change and Africa. Oxford Review of Economic Policy, (24), 337-353.

Collins, M., R. Knutti, J. Arblaster, J.-L. Dufresne, T. Fichefet, P. Friedlingstein, X. Gao, W.J. Gutowski, T. Johns, G. Krinner, M. Shongwe, C. Tebaldi, A.J. Weaver & M. Wehner, 2013, Long-term Climate Change: Projections, Commitments, and Irreversibility. In: Climate Change 2013: The Physical Science Basis. Contribution of Working Group I to the Fifth Assessment Report of the Intergovernmental Panel on Climate Change [Stocker, T.F., D. Qin, G.-K. Plattner, M. Tignor, S.K. Allen, J. Boschung, A. Nauels, Y. Xia, V. Bex & P.M. Midgley (Eds.)]. Cambridge University Press, Cambridge, United Kingdom & New York, NY, USA.

Eisenack, K. & Stecker, R., 2013, A Framework for Analysing Climate Change Adaptations as Actions. Mitigation and Adaptation Strategies to Global Change, University of Oldenburg, 22.

Ellen, R. & Harris, H., 1996, Concepts of Indigenous Environmental Knowledge in Scientific and Development Studies Literature–A Critical Assessment. Draft Paper Presented at East-West Environmental Linkages Network workshop 3. Canterbury.

Elum, Z.A., Madise, D.M. & Marr, A., 2017, Farmer's Perception of Climate Change and Response Strategies in three Selected Provinces of South Africa. Climate Risk Management, (16) 246–257.

Emgeru, A., 2012, Role of Indigenous Knowledge in Climate Change Adaptation: A Case Study of the Teso Sub-Region, Eastern Uganda. Indian Journal of Traditional Knowledge, 11 (2), 217-224.

Fernández-Llamazares, Á., Díaz-Reviriego, I., Luz, A.C., Cabeza, M., Pyhälä, A., & Reyes-Garcia, V., 2015, Rapid Ecosystem Change Challenges the Adaptive Capacity of Local Environmental Knowledge. Global Environmental Change (31) 272–284.

Food and Agriculture Organisation, FAO, 2013, Climate-Smart Agriculture Sourcebook, Rome, 5-8.

Food and Agriculture Organisation, FAO, 2014, Definitional Framework of Food Loss-Global Initiative on Food Loss and Waste Reduction. Working Paper, Rome.

Gyampoh, B. A., Amisah, S., Idinoba, M. & Nkem, J., 2009, Using Traditional Knowledge to Cope with Climate Change in Rural Ghana. Unasylva (English Ed.), (60), 70-74.

Heltberg, R., Siegel, P.B. & Jorgensen, S.L., 2009, Addressing Human Vulnerability to Climate Change: Toward a 'no-regrets' approach. Global Environmental Change, (19), 89-99.

Hiwasaki, L., Luna, E., Syamsidik, & Shaw, R., 2014, Local and Indigenous Knowledge for Community Resilience: Hydro-meteorological Disaster Risk Reduction and Climate Change Adaptation in Coastal and Small Island Communities. Jakarta, UNESCO, 60.

Hulme, M., Doherty, R., Ngara, T., New, M. & Lister, D., 2001, African Climate Change: 1900-2100. Climate Research, (17), 145-168.

Intergovernmental Panel on Climate Change, IPCC, 2001, Annex B. Glossary of Terms. Third Assessment Report (TAR). IPCC Secretariat, Bonn, 365-388.

Intergovernmental Panel on Climate Change, IPCC, 2007, Climate Change 2007: Impacts, Adaptation and Vulnerability. Contribution of Working Group II to the Fourth Assessment Report of the Intergovernmental Panel on Climate Change. Cambridge University Press, Cambridge UK, 976.

Intergovernmental Panel on Climate Change, IPCC, 2014, Climate Change 2014: Impacts, Adaptation and Vulnerability, Contribution of Working Group II to the IPCC Fifth Assessment Report. Cambridge University Press, Cambridge, United Kingdom & New York, 1774.

International Geosphere-Biosphere Programme, IGBP, 2008, Global Change News Letter, No. 72, December 2008. IGBP Secretariat, Stockholm, 28.

International Institute for Environment and Development, IIED, 2015a, Vulnerable Communities: Getting their Needs and Knowledge into Climate Policy. IIED Briefing, November 2015, London, 1-4.

International Institute for Environment and Development, IIED, 2015b, Vulnerable Communities: Climate Adaptation that Works for the Poor. IIED Briefing, November 2015, London, 1-4.

International Labour Organisation, ILO, 2017, Indigenous Peoples and Climate Change: From Victims to Change Agents through Decent Work. ILO Office, Geneva, 45.

Kolawole, D.O., P., Wolski, B., Ngwenya & G., Mmopelwa, 2014, Ethno- meteorology and Scientific Weather Forecasting: Small Farmers and Scientists' Perspectives on Climate Variability in the Okavango Delta, Botswana. Climate Risk Management (4-5), 43-58.

Krishna, V., 2011, Indigenous Communities and Climate Change Policy: An Inclusive Approach. The Economic, Social and Political Elements of Climate Change, Springer.

Macchi, M., Boedhihartono, A., Wolfangel, C., Howell, M. Oviedo, G., Gotheil, S., & Cross, K., 2008, Indigenous and Traditional Peoples and Climate Change. International Union for Conservation of Nature, 9-22.

Manatsa, D., Unganai, L., Gadzirai, C. and Behera, S.K., 2012, An Innovative Tailored Seasonal Rainfall Forecasting Production in Zimbabwe. Natural Hazards (64), 1187–1207.

Manjengwa, J., 2012, Environment, Natural Resources and Poverty Reduction: Capturing and Sharing the Gains of Natural Resources Exploitation. In: Understanding poverty, promoting wellbeing and sustainable development A sample survey of 16 districts of Zimbabwe. Harare: Sable Press, 79-94.

Mararike, C., 2014, Land: An Empowerment Asset for Africa: The Human factor approach. Harare: University of Zimbabwe Publishers, 83-102.

Mason, S.J., Joubert, A. M., Cosijn, C. and Crimp, S. J., 1996, Review of seasonal forecasting techniques and their capabilities. Water SA 22 (3).

Mathews, T., Mullan, D., Wilby, R.L., Broderick, C., & Murphy, C., 2014, The Household Economy Approach: Managing the Impact of Climate Change on Poverty and Food Security in Developing Countries. Climate Risk Management, (4-5), 59-68.

Miller, N.P., Dow, K., Tuler, S., Webler, T., Whitehead, J., & Miller, K.M., 2014, Integrating Scientific and Local Knowledge to inform Risk-based Management Approaches for Climate Adaptation. Climate Risk Management, (4-5), 17-31.

Mwaura, P., 2008, Indigenous Knowledge in Disaster Management in Africa. UNEP, Nairobi, 21-33.

Nakashima, D.J., Galloway McLean, K., Thulstrup, H.D., Ramos Castillo, A. & Rubis, J.T., 2012, Weathering Uncertainty: Traditional Knowledge for Climate Change Assessment and Adaptation. Paris, UNESCO & Darwin, UNU, 120.

Odero, K., 2011, The Role of Indigenous Knowledge in Responding to Climate Change: Local-global Perspectives. Panel 10: Roles of Local and Indigenous Knowledge in Addressing Climate Change. African Adapt, Climate Change Symposium 2011.

Omambia, C.S., Nyatichi, A. & Gu, Y., 2010, The Cost of Climate Change in Tanzania: Impacts and Adaptations. Journal of American Science, (6), 182-196.

Oppenheimer, M., M. Campos, R. Warren, J. Birkmann, G. Luber, B. O'Neill, & K. Takahashi, 2014, Emergent risks and key vulnerabilities. In: Climate Change 2014: Impacts, Adaptation, and Vulnerability. Part A: Global and Sectoral Aspects. Contribution of Working Group II to the Fifth Assessment Report of the Intergovernmental Panel on Climate Change [Field, C.B., V.R. Barros, D.J. Dokken, K.J. Mach, M.D. Mastrandrea, T.E. Bilir, M. Chatterjee, K.L. Ebi, Y.O. Estrada, R.C. Genova, B. Girma, E.S. Kissel, A.N. Levy, S. MacCracken, P.R. Mastrandrea, & L.L. White (Eds.)]. Cambridge University Press, Cambridge, United Kingdom & New York, NY, USA, 1039-1099.

Organisation for Economic Co-operation and Development, OECD, 2009, Integrating Climate Change Adaptation into Development Co-operation, Policy Guidance. OECD, Paris, 193.

Peloquin, C. & Berkes, F., 2009, Local Knowledge, Subsistence Harvests and Social–Ecological Complexity in James Bay. Human Ecology, (37), 533-545.

Philander, S.G., 2008, Encyclopedia of Global Warming and Climate Change (1-3), SAGE Publications, Los Angeles.

Rhodes, E.R., Jalloh, A. & Diouf, A., 2014, Review of Research and Policies for Climate Change Adaptation in the Agriculture Sector in West Africa. Future Agricultures, Working Paper 090, 12-18.

Rose, C., Dade, P., Gallie, N., & Scott, J., 2005, Climate Change Communications–Dipping a Toe into Public Motivation. Available from <http://www.campaignstrategy.org/ valuesvoters/climatechange communications.pdf> [Accessed on 18 January 2017].

Rosenzweig, C., Iglesias, A., Yang, X., Epstein, P.R. & Chivian, E., 2001, Climate Change and Extreme Weather Events; Implications for Food Production, Plant Diseases and Pests. Global Change & Human Health, (2), 90-104.

Scoones, I., 1998, Sustainable Rural Livelihoods: A Framework for Analysis, Institute of Development Studies, Brighton.

Seddon, N., Hou-Jones, X. Pye, T., Reid, H., Roe, D., Mountain, D. & Rizvi, A.R., 2016, Ecosystem-based Adaptation: A win–win Formula for Sustainability in a Warming World? International Institute for Environment and Development (IIED) Briefing, London, 1-4.

Seddon, N., Reid, H., Barrow, E., Hicks, C., Hou-Jones, X., Kapos, V., Rizvi, A.R., & Roe, D., 2016, Ecosystem-based Approaches to Adaptation: Strengthening the Evidence and Informing Policy, Research Overview and Overarching Questions. International Institute for Environment and Development (IIED), London, 1-20.

Sen, B., 2005, Indigenous Knowledge for Development: Bringing Research and Practice Together. The International Information & Library Review, (37), 375-382.

Senanayake, S., 2006, Indigenous Knowledge as a Key to Sustainable Development. Journal of Agricultural Sciences, (2), 87-94.

Sivakumar, M., Das, H. & Brunini, O., 2005, Impacts of Present and Future Climate Variability and Change on Agriculture and Forestry in the Arid and Semi-arid Tropics. Climatic Change, (70), 31-72.

Smit, B., & Pilifosova, O., 2001, Adaptation to Climate Change in the Context of Sustainable Development and Equity. Climate Change 2001: Impacts, Adaptation and Vulnerability, Cambridge: Cambridge University Press, 879-905.

Snow, M. & Prasad, D., 2011, Climate Change for Building Designs: An Introduction. Environment Design Guide (66), 3.

Stern, N., 2007, Stern Review: The Economics of Climate Change. Cambridge and New York: Cambridge University Press, 404-413, 430- 443.

Swiderska, K. & Malmer, P., 2016, Climate Change and Biocultural Adaptation in Mountain Communities: Second International Learning Exchange of the International Network of Mountain Indigenous People (INMIP). International Institute for Environment and Development, 46.

Tanyanyiwa, V.I. and Chikwanha, M, The role of indigenous knowledge systems in the management of forest resources in Mugabe area, Masvingo, Zimbabwe. Journal of Sustainable Development in Africa. 3(3). 132-149

Thompson, J. & Scoones, I., 1994, Challenging the Populist Perspective: Rural People's Knowledge, Agricultural Research and Extension Practice. Agriculture and Human Values, (11), 58-76.

Töpfer, K. & Hunter, J.W., 2002, Climate Change Information Kit. UNEP & UNFCCC, UNFCCC Secretariat, Bonn.

Toulmin, C. & Huq, S., 2006, Sustainable Development: Africa and Climate Change. International Institute for Environment and Development (IIED), London, 1-2.

Toulmin, C., Huq, S. & Rockström, J., 2005, Sustainable Development: Africa and Climate Change. International Institute for Environment and Development (IIED), London, 1-2.

United Nations Environment Programme, UNEP, 2010, Making the Case for Ecosystem-based Adaptation: Building Resilience to Climate Change. UNEP, Nairobi, 11.

United Nations Framework Convention on Climate Change, UNFCCC, 2011, The Nairobi Work Program on Impacts, Vulnerability and Adaptation to Climate Change; Climate Change and Freshwater Resources: a Synthesis of Adaptation actions undertaken by Nairobi work program partner organizations. UNFCCC Secretariat, Bonn, 10- 11.

United Nations Framework Convention on Climate Change, UNFCCC, 2006, UNFCCC Handbook. UNFCCC Secretariat, Bonn.

United Nations Framework Convention on Climate Change, UNFCCC, 2006, UNFCCC Handbook. UNFCCC Secretariat, Bonn.

Waha, K., Müller, C., Bondeau, A., Dietrich, J.P., Kurukulasuriya, P., Heinkef, J.,Lotze-Campen H., 2013, Adaptation to Climate Change through the Choice of Cropping System and Sowing Date in Sub- Saharan Africa. Global Environmental Change (23) 130–143.

Ward, B. & Menezes, S., 2008, Communicating on Climate Change: An Essential Resource for Journalists, Scientists and Educators. Metcalf Institute for Marine & Environmental Reporting, Narragansett, 27.

Wong-Parodi, G., Fischoff, B. & Strauss, B., 2015, Resilience vs Adaptation: Framing and Action. Climate Risk Management (10), 1-7.

World Meteorological Organization, 2010, Guide to Agricultural Meteorological Practices, World Meteorological Organization, Geneva.

Yanda, P.Z. & Mubaya, C.P., 2011, Managing a Changing Climate in Africa: Local Level Vulnerabilities and Adaptation Experiences. African Books Collective.

Ziervogel, G., & Calder, 2003, Climate Variability and Rural Livelihoods: Assessing the Impact of Seasonal Climate Forecasts in Lesotho. Area 35 (4), 403– 417.

Chapter Two

CHANGE AGENTS IN CLIMATE CHANGE MITIGATION-THE ROLE OF ZIMBABWE'S SOCIAL WORKERS

Tatenda Goodman Nhapi, *Social Worker, UK Local Authority*

Takudzwa Leonard Mathende, *Social Worker, UK Local Authority*

Abstract

This chapter is grounded in historiography that analyses Zimbabwean empirical literature and policy documents in terms of opportunities and platforms. Rural Development-oriented frontline social workers can contribute to the mitigation of human livelihoods actions that eat away the very fragile environment the same livelihoods depend on. Grounded in ideals of Sustainable Development Goals, the article harnesses a Person in Environment approach for arguing social workers' active role in the climate change mitigation agenda. This would be to broker harmony of rural communities harnessing their natural resources whilst mitigating climate change impacts. Discussion of developmental, social work and indigenisation of social work in Africa is inadequate without focusing on environmental sustainability. In the African context, social workers' methods of intervention target poor communities, who in harnessing natural resources end up as one of the key drivers of environment degradation based on poverty as the biggest polluter assertion. Hurungwe District in Mashonaland West, Zimbabwe is a profitable tobacco farming region affected by adverse climate change impacts due to tree cutting for tobacco. Recommendations propose that social workers should collaborate in environmental justice advocacy alongside other Rural Development (RD) intervention initiatives. Indigenous Knowledge Systems approaches can harness and enrich climate change awareness interventions in relation to Zimbabwe's rural development programmes.

Key words: Change Agents • Climate Change • Mitigation • Social Workers • Zimbabwe

Introduction

This chapter examines the interface between rural development (RD) and social work using the lens of climate change adaptability. The Government in pursuing an enhanced transformative climate response trajectory, through scaling up climate adaptation and mitigation actions, need the expertise of professionals such as social workers. African social scientists usually in collaboration with scholars from other continents, have begun to respond to the climate crisis, focusing particularly on its implications on various facets of development and livelihoods (Murombedzi, 2014). Like other developing countries, Zimbabwe's experiences in unprecedented climate variability and change impacts are mainly felt by vulnerable poor communities, women, and children. With an economy founded on sectors vulnerable to climatic change, namely agriculture, forestry, energy, tourism, and industry, among others, the impacts from climate change have a significant effect on access to food, water, health, settlements, and energy (Dube and Phiri, 2013).

Existing climate data reveals a worsening scenario if significant action is not taken. Seventy percent of the sub-Saharan Africa population lives in rural areas relying largely on working small farms for their sustenance (Matondi and Rukuni, 2010). Zimbabwe has an abundant and diversified natural resource base consisting of both land and water resources including minerals. According to Chasi in Manjengwa (2012), on a global scale, Zimbabwe has one of the highest endowments of natural resources. The forestry resources cover approximately 66% of the total land area (United Nations Country Team- UNCT, 2014). As UNCT further note, 62.6% of Zimbabwe households are poor as displayed by the per capita consumption expenditures below the Total Consumption Poverty Line (TCPL). Of these poor households, 76% live in rural areas compared to 38.2% in urban areas (ZimStats, 2011).

Zimbabwe has an enabling environmental protection legislative framework principally under the Environmental Management Act (Chapter 20:27), administered by the country's statutory arm Environmental Management Agency (EMA) which falls under the Ministry of Environment, Water and Climate. The Act's Chapter six, section 97 criminalizes failure for the undertaking of a mandatory Environmental Impact Assessment (EIA). It asserts that EIA needs to be commissioned prior to any major developmental project (prescribed activities) to guarantee the mitigation of social and ecological impacts of activities such as mining.

Imperatives for adaptation are clear as Zimbabwe experiences climate change effects such as rainfall variability, extreme events and warming trends increasingly endangering rain-fed agriculture, the economy and the poor's livelihoods (Bhatasara and Mandizadza, 2014). Climate change's negative effects are threatening to reverse development gains in many parts of the world especially in Sub-Saharan Africa (Dube and Phiri, 2013). It is now an accepted scientific phenomenon that the global climate is changing. Essentially, rural farmers in marginalised ecosystems, need to adapt to climate change, yet farming systems adaptation should not be considered solely as a contemporary phenomenon or a historical one (Bhatasara and Mandizadza, 2014). The Intercontinental Panel on Climate Change (IPCC) Third Assessment Report and Zimbabwe's INC suggest that by 2050 temperatures and rainfall over the country will be 2-4°C higher and 10-20% less than the 1961-90 baseline respectively. Model simulations show annual rainfall declining by 5-20% of the 1961-90 average by 2080 in all the country's major river basins (GEF, 2013).

Consequently, this chapter argues for tailor-made climate mitigation interventions in RD endeavours. These initiatives should be aided by harnessing community development core competencies possessed by social workers. The framing of the social work profession scope indicates how it addresses the barriers, inequities, and injustices existing in society (Mugumbate and Maushe, 2014). Furthermore, as Kangethe (2014) observes, social work is an important humanistic profession recognized as a respectable discipline addressing different vulnerabilities that people face. Again, Chitereka (2009) notes, poverty manifestation in the form of majority spending less than a dollar a day on their livelihood, degraded environment, and homelessness, is increasing alarmingly in Africa. This is mainly caused by political regimes which do not care for the welfare of their citizens.

Methodology

This work is based on a historiographical analysis of empirical literature and policy documents. Rural Development oriented frontline social workers can contribute to the mitigation of human livelihoods actions that eat away the very fragile environment the same livelihoods depend on. The case of Hurungwe District, located 350km, Northwest of Harare is discussed. Hurungwe lies in a remote, rural part of Mashonaland West Province, one of the 10 geopolitical provinces. It is adjacent to Mana Pools National Park and covers 150,018ha. Hurungwe is the second largest of Zimbabwe's 56 districts. It is located in Highveld's prime cereal and cash crop resettlement agricultural zone based on the rain-fed agricultural production of both cash and food crops. Additionally, Hurungwe Rural District Council has for the second time accused the Forestry Company of Zimbabwe (FCZ) of facilitating deforestation by unprocedurally issuing wood permits to tobacco companies working in the district. The Tobacco Industry Marketing Board (TIMB) contends that the number of registered tobacco farmers has grown to 75,580 and of these 24,455 are in Hurungwe arguably the highest number. The communal subsistence farming model A1 (either as a *villagised* or self-contained model variant) has been one of the success stories of Zimbabwe's climate change adaptation. Thus, this chapter is largely reflective in its approach, referring to various Zimbabwean studies and survey reports from a variety of institutions. Data were largely collected by documentary search, and content analysis is facilitating specific conclusions.

Controversies, Issues, and Problems

The environment is taking centre stage in local, national and global discourse and policies (Murombedzi, 2014). In the Marxist discourse on environmental change usage, Foster (1999) notes Marx's metabolic concept usage to capture material breakup of human beings in capitalist society from the natural conditions of their existence. To argue that large-scale capitalist agriculture created such a metabolic rift between human beings and the soil was to argue that basic conditions of sustainability had been violated (Foster, 1999). As Foster further notes in 'Capitalist production', Marx (1861–63, 1971b, 301) wrote, 'turns toward the land only after its influence has exhausted it and after it has devastated its natural qualities'.

Furthermore, climate change has come to dominate contemporary environmental debates and the enhancement of development policy (Murombedzi, 2014). In 2012, University of Zimbabwe's(UZ) Institute of Environmental Studies(IES) surveyed 12 districts in Zimbabwe on a study titled *Moving Zimbabwe Forward: An Evidence Based Policy Dialogue Initiative,* The IES team concluded that poverty was high especially in rural areas. Eight-one percent of the sampled households lived in poverty. Households living in the dry and marginal lands of Agro-Ecological Regions IV and V (characterised by low rainfall less than 250mm and high mean temperatures, above 19^0 C) lived in poverty or extreme poverty. Promoting access to alternative livelihood options across these different areas is critical (Chimhowu, Manjengwa, and Feresu, 2010). An important indicator of a livelihood system's status is its ability to provide individual households with enough income to meet their basic needs. The poor should be able to sustainably manage and benefit from rich and abundant natural resources to address poverty and promote inclusive growth in Zimbabwe (Manjengwa, 2012). Furthermore, the Human Factor approach usage is relevant for discourse analysis in this chapter. The main claim of the Human Factor approach as propounded by Mararike (2014) is that no nation or country can sustain its development activities without people who are patriotic, reliable, committed and disciplined. For this role, social workers are fit for this purpose as change agents with core competencies to make tailor-made rural development programmes. Indigenous Knowledge Systems (IKS) need to be harnessed comprehensively to facilitate social workers led pro-poor rural development interventions that mitigate climate change. IKS consists of a body of knowledge, beliefs, traditions, practices, and institutions developed and sustained by indigenous, peasant and local communities in their interaction with the biophysical environment (Bhatasara and Mandizadza, 2014). Through diagnostic rather than prognostic, indigenous indicators, such as trees physiological behaviour, behaviour of frogs and sounds of birds, are related to the response of certain animals and plants to the already prevailing weather form. Whilst IKS has been a source of resilience; climate change provokes serious threats to both livelihoods and indigenous climate knowledge (Bhatasara and Mandizadza, 2014).

NGOs are often guilty of by-passing rural district councils, undermining development planning system (International Institute for Environment and Development (IIED), 2000). NGO proliferation creates activities that make it difficult to coordinate rural livelihoods initiatives. Their planning approaches vary according to the scale and scope of activities and their attitude towards development endeavours and overall objectives (IIED, 2000).

RD Dynamics & Contexts, Literature Review and Theoretical Framework

Menes in Mupuva (2015) notes RD as a process of quality life and economic well-being of people improvement in relatively isolated and sparsely populated areas. Traditionally, according to Mapuva, rural areas' development centred on agriculture and forestry but due to globalization and diversification needs, rural development has transformed to mining, tourism, and manufacturing. Rural development also caters for locally produced economic development strategies and rural development programmes. As a result, there has been a shift from a top-down to bottom-up approach by NGOs (Mapuva, 2015).

While several studies on climate change impact on local livelihoods exist there is, however, paucity of literature dedicated to Zimbabwe (Dube and Phiri, 2013). Five natural farming regions based mainly on rainfall and temperature, seven land types based on land tenure make up Zimbabwe. The majority of Zimbabwe's population live in communal areas (CAs) (previously known as African reserves or Tribal Trust Lands). These communal areas lie in marginal agricultural areas (IIED, 2000). Furthermore, Zimbabwe's rural economy is strongly influenced by ecology, land tenure and use, population density and land distribution (FAO Documents Repository, n.d). Zimbabwe's rural livelihoods like in most Sub-Saharan countries revolve around rain-fed farming (Dube and Phiri, 2013).

'Social characteristics' of rural areas include demographic structure, ethno-linguistic characteristics, social structure, inheritance systems, religious and cultural beliefs and practices as well as individual and group attitudes. Key rural poverty drivers have been low agricultural productivity, aggravated by demographic growth, motivating migration to urban centres. Other challenges to environmental sustainability in rural areas include highly sectoral and compartmentalized understanding of development issues, competing demands of economic and industrial development, resource conservation, pollution control and capacity of Government (Bhatasara and Mandizadza, 2014). From 1990 to 2010, Zimbabwe was in the top ten countries with the largest annual net loss of forest area (UNCT, 2014). Agenda 21, the global plan of action adopted at the 1992 Earth Summit in Rio de Janeiro, claims that desertification affects about one-sixth of the world's population, 70 percent of all drylands, amounting to 3.6 billion hectares, and one-quarter of the total land area of the world (Leach 1999). Also, in the Matobo District, Matabeleland South Province, there are increasing temperatures and decreasing precipitation, thus, altering the natural environment and leading to the disappearance of natural habitat, flora, and fauna (Phiri and Dube (2012). This has critical implications on rural livelihoods.

According to IIED (2000), historically top-down, blueprint philosophies dominated RD interventions planning approaches until the mid-1980s. During this epoch, bottom-up and participatory planning was emphasised. In the 1990s emphasis was on planning processes that were much more open-ended and dynamic. Zimbabwe's rural areas have experienced land alienation due to a settler system that took the most productive land for itself and relegated the black majority to the marginal agricultural areas (IIED, 2000). Zimbabwe has a long history of agricultural extension and farmers are generally knowledgeable of good conservation measures (Manjengwa, 2012). According to IIED (2000), Government-led rural development programmes are as follows: Land and environment; Land-use and agriculture; Economic non-farming; Infrastructure; Settlement; Poverty alleviation and drought mitigation; Institutional support to rural planning. Dube and Phiri (2013) note that in trying to understand emerging trends of climate change and its effects in local ecologies, it is important to address the negative effects of climate change.

Results and Discussion

Climate Change, RD and role of Social Workers in Hurungwe

There are a number of NGOs operating in Hurungwe, and some of these include Lead Trust, World Vision Hurungwe Area Development Programme, Amanzi Trust and Goal Zimbabwe. These NGOs employ social workers in some of their ranks and programmes. The project developer, Carbon Green Africa has local allies in Hurungwe that are farmers and the Hurungwe Rural District Council (HRDC). A range of benefits are offered in the form of carbon dividends and 'alternative livelihood' projects in exchange for protecting forests and planting trees (Scoones, 2015). As the notional 'traditional' and 'administrative' owners of the land, Scoones observes they should have the authority but are pitched against powerful forces with other ideas about resource use and economic priorities. These include politically-connected tobacco farmers who migrated into the area in the 1980s and 90s at the invitation of the same local Korekore leaders who are now backing carbon dividends (Scoones, 2015). The chapter contends that social workers are think tanks and bodies and as NASW should be at the forefront of commissioning participatory action research into initiatives such as Kariba Reduced Emissions for Deforestation and Forest Degradation (REDD+)[1] project in Hurungwe. In a report by Matonho (2015), one rural farmer, a participant in REDD project living in densely-forested zones conducive for REDD+ projects indicated waiting for ages for promised returns by private sector players and NGOs from protecting forests and woodlands. Matonho further notes that at the Lima Conference held in December 2016 Washington Zhataka, a senior official for the Ministry of Environment, Water and Climate, told delegates at the UNFCCC COP 20 report back meeting that despite the hype given to the Redd+ project, rural district councils (RDCs) could get as little as $50,000 while an investor in the project walks away with millions of dollars. Another farmer, Moses Mutale from Nyaminyami District in Kariba, is quoted by Matonho as indicating that 'REDD+ implementers need to be transparent in the signing of contracts with communities and assist them with responsive community-driven alternative livelihood projects, and save the remaining forests and woodlands'.

A 2016 Hurungwe report by Pito (2016) indicates that Hurungwe Rural District Council has for the second time accused the Forestry Company of Zimbabwe (FCZ) of facilitating deforestation by unprocedurally issuing wood permits to tobacco companies working in the district. Although the RDC is involved in an increasing range of RD programmes (RDPs), their lack of financial resources means that they are unable to offer many independent programmes (IIED, 2000). Furthermore, for the 2015 season, TIMB noted that

---

[1] Deforestation and land degradation globally contribute significantly to carbon emissions, and addressing these has become a major policy priority. Carbon offset approaches, mediated by carbon markets and facilitated by international accords and global climate finance, have become especially popular. In such schemes carbon emissions in one part of the world (usually the industrialised north) are offset by initiatives that reduce emissions in another part of the world where there are plentiful forests, and opportunities for new carbon sequestration (such as Africa). Such projects can, it is argued, additionally focus on poverty reduction and biodiversity protection, creating a 'win-win' scenario, rather than a feared 'green grabbing'.

the number of registered tobacco farmers has grown as indicated earlier on. With Hurungwe having the biggest number of registered growers. Observers have noted that the A1 farming model has been successful to a large extent in climate change adaptation (Muzulu, 2014).

Pito (2016) reports that paramount Chief Chundu of Chundu in Hurungwe indicated some contract farmers in his area are so insolent that they cut down sacred trees despite him having warned them against such acts on several occasions '*These contract farmers are defying my orders; they even cut down sacred trees under which we make rain-making prayers, all in the name of growing tobacco. No wonder why we have devastating droughts. When the spirits go angry droughts, and unusual happenings besiege us*'. The new social, cultural and economic landscape, evolving through waves of migration, is one such scenario where a simple REDD project is immensely difficult to implement, as divisions based on ethnicity, class, gender, economic priority and more divide 'the community' that is notionally involved in the project. Greater social development paradigm adoption by African social workers results in people being more self-conscious and aware of their rights thus minimising poor governance as the majority participate in issues affecting their general welfare (Chitereka, 2009). Furthermore, people in rural agrarian economies have often survived harsh and changing climatic conditions. In the process, they develop imaginative adaptation measures that build resilient livelihoods (Bhatasara and Mandizadza, 2014). Finally, one of the main concerns among those involved in the social aspects of rural development is the impact of politics on distributional issues-that is, who gains and who loses as a result of development efforts. There is a close relationship between political power and influence on the one hand and the level of economic and social well-being on the other (FAO Documents Repository, n.d).

Conclusion and Policy Options

The Social Work role in RD and the climate change mitigation agenda is critical. African Social work embraced the curative or remedial approach to solving social problems during its inception (Chitereka, 2009). Notably, the majority of social workers employed by NGOs are involved in developmental, social work, with a specific bias to rural areas. The main role of the social workers is to mobilise communities to take an active interest in the problems affecting them (Chitereka, 2009). The International Federation of Social Workers (IFSW), the International Association of Schools of Social Work (IASSW), and the International Council on Social Welfare (ICSW)'s Global Social Work Agenda 2012-2016, note that people's health and wellbeing suffer as a result of inequalities and unsustainable environments related to climate change, pollutants, war, natural disasters and violence to which there are inadequate international responses. Jarvis (2009) notes that social work institutions have academics who have been producing professionals who lack the capacity to address the nexus of environmental justice, social work, and social justice.

Most rural societies are relatively 'integrated'. The various components of agricultural and non-agricultural life that include 'economic', 'social', 'political', religious and secular life are closely interrelated. Rural people do not easily recognize the role of planners, extension workers, and other government officials'. For them, it is rather difficult to make a distinction between responsibilities of different agencies or 'economic' and 'social' planning because in their own lives all these things are interdependent (FAO Documents Repository, n.d; Kujinga *et al.*, 2012). A study in the Chimanimani District of Zimbabwe analysed the interface between academic researchers, development practitioners, district institutional actors and local communities in a project aimed at forest resources management enhancement. Results indicated that for the involvement and accommodation of various actors' interests, quantitative and qualitative action research strategies and participatory geographic information systems approach are critical for effective environmental management.

Local and external actors coalesce despite their social differences usually by defining social interfaces as technical expertise, material and financial resources possessed by external actors. External actors, despite their technical and financial resources, lack knowledge of the local area. Actors such as local authorities, government departments, quasi-government organisations, officials and local communities might lack certain technical, financial and material resources, but have enormous knowledge about their areas, and are passive recipients of the packages that can come with external actors (Kujinga *et al.*, 2012). Based on Kujinga *et al.*, (2012), social workers in rural development and climate change interventions can bridge the gap and harness synergies between various actors for desired outcomes of climate change adaptation and mitigation.

Questionnaires, case studies, empiricism, philosophy, and interviews are now redundant research techniques not relevant for social workers' usage (Mbigi, 2014). Currently, participation action research discourse analysis, phenomenology, critical theory, cooperative enquiry, grounded theory, appreciative inquiry and critical rationalism are alternative viable research approaches. Social workers have facilitated participatory research particularly youth and women organizations in cooperation with Rural District Councils and resource persons to develop and adapt to climate change practices. Moreover, community leaders and volunteers in a given area would be capacitated to act as champions for climate change adaptation, including addressing issues to do with gender balance.

The Zimbabwean Social worker representative body, National Association of Social Workers (NASW), could seek funding to commission participatory action research in Hurungwe to establish climate change impacts on rural development from a social development perspective. This calls for Schools of Social Work in Zimbabwean universities to establish within their ranks, Centres for Applied Social Work. The call for the establishment of these is getting louder. Such an action research-oriented institution could harness synergies with established sister institution of applied and Environmental Sciences such as the Centre for Applied Social Sciences(CASS) and the Institute of Environmental Studies (IES) both at the University of Zimbabwe for drafting concept notes to seek funding for commissioning of action research. This also facilitates further knowledge procreation as both undergraduate, and masters' students become fieldworkers who collect data in such research studies. The idea is to galvanise their action research conduct capabilities with a bias towards rural development and climate change knowledge management.

The social work curriculum is based on both theory and practice; thus the provision of the sound theory is as critical as fieldwork experience. It is critically important for social work students to acquire, in a classroom environment, practice principles, values, and ethics and the scientific basis for practice and theoretical content application to real-life situations to become professional social workers (Dhemba, 2012). Since the mid-1980s governments have been pressured to hand over responsibilities to local governments, non-government organizations and people's organizations through a decentralisation initiative. In the 1990s Zimbabwe abandoned socialist command economy in favour of a liberalised market-led economy. This process led to an integrated and participatory planning system, which was responsive to people's needs and priorities like environmental sustainability and Local Agenda 21implementation. A new form of environmental planning involving EIAs for all new developments and district environmental action plans in selected communal areas was mooted. This was meant to 'de-politicize' the development process through changing the importance of the various political actors and channels. Kaseke in Chogugudza (2009) acknowledges that African social work education and training emphasis has fundamentally shifted to a social development approach placing greater significance on macro intervention to reach the individual or group.

Zimbabwean RD oriented social workers should infer on the Global Agenda for Social Work, in the context of climate change interventions which stipulates that social work organisations should promote standards in education and practice that facilitate sustainable social development outcomes. These outcomes include the prevention, mitigation, and response to disasters through sound research. Community-based natural product enterprises and commercial ventures seeking supplying markets with value-added natural products bring in greater benefits to communities that manage and use natural resources through a vibrant approach that should be harmonised led by RD social workers' interventions that include climate change mitigation outcomes. Examples of such initiatives in Zimbabwe include Amacimbi Development Association in Gwanda which recorded a return on investment of 166%; Creative Oils Investments, in Rushinga attained self-sufficiency within the first seven months of operation, after repaying a loan of USD 700 for machinery and working capital (Manjengwa, 2012).

The Ministry of Environment Water and Climate Management's slogan is *'The environment is everybody's responsibility'* and one of the principles of the Environmental Management Act is that development must be socially, environmentally and economically sustainable (Manjengwa, 2012). GEF found, in a study in Chiredzi in south-east Zimbabwe, that bottom-up project design and participatory processes are crucial for strong ownership and identification of adaptation responses that are acceptable in a local and cultural context. Community participatory climate risk analysis made it possible to generate more information on the temporal and spatial dynamics of drought and its impacts on the project area than could be obtained from modelling.

As a final point, in order to enhance district and community levels preparedness to climate-sensitive disaster risk management, plans need to be in place particularly for infrastructure design and human settlements planning. The challenges reinforce the need to establish coherence between disaster risk reduction (DRR), climate change adaptation and gender sensitivity in the planning of development interventions. Zimbabwe should promote policies that include climate change education and research to ensure the development of a resilient population and competent research centres. The government should also increase provision and uptake of unbiased and credible evidence in policy processes to strengthen adaptation strategies of vulnerable communities through employing social workers.

References

Bhatasara, S. & Mandizadza, E. J., 2014, Climate Change and Indigenous Knowledge in Zimbabwe: A human factor approach. In: C. Mararike, (Ed., 2014), Land: An Empowerment Asset for Africa: The Human factor approach. Harare: University of Zimbabwe Publishers, 83-102.

Chimhowu, A., Manjengwa, J., and Feresu, S. 2010. Moving Forward in Zimbabwe: Reducing Poverty and Promoting Growth. Institute of Environmental Studies / Brooks World Poverty Institute (BWPI), Harare.

Chitereka, C., 2009, Social Work Practice in a Developing Continent: The case of Africa. Advances in Social Work, 10(2), 144-156.

Chogugudza, C., 2009, Social Work Education, Training and Employment in Africa: The case of Zimbabwe. Ufahamu: A Journal of African Studies, 35 (1), 2-7.

Dhemba, J., 2012, Fieldwork in Social Work Education and Training: Issues and Challenges in the Case of Eastern and Southern Africa. Social Work and Society, 10 (1).

Dube, T. & Phiri, K., 2013, Rural Livelihoods Under Stress: The Impact of Climate Change on Livelihoods in South Western Zimbabwe. American International Journal of Contemporary Research, 3 (5), 11-25.

Food and Agricultural Organisation Documents Repository, n.d Guidelines on social analysis for rural area development planning. [Online] Available at: http://www.fao.org/docrep/T1660E/t1660e02.htm# P709_86169 [Accessed 20 June 2016].

Foster, J. B., 1999, Marx's Theory of Metabolic Rift: Classical Foundations for Environmental Sociology. American Journal of Sociology, 366-405.

International Institute for Environment and Development-IIED, 2000, Local Strategic Planning and Sustainable Rural Livelihoods Rural District Planning in Zimbabwe: A case study. [Online] Available at: http://pubs.iied. org /pdfs /7827IIED.pdf [Accessed 28 May 2016].

International Federation of Social Workers, 2012, The Global Agenda for Social Work and Social Development A call to action. [Online] Available at: http://cdn.ifsw.org/assets/globalagenda2012.pdf. [Accessed 6 May 2016].

Jarvis, D., 2009, Environmental Justice and Social Work: A Call to Expand the Social Work Profession to Include Environmental Justice. Columbia Social Work Review (36).

Kujinga, K., Chingarande, S., Mugabe, P. & Nyelele, C., 2012, Interface Between Research, Development and Local Actors in Enhancing Sustainable Forest Resources Management: Lessons from Chimanimani District, Zimbabwe. Journal of Social Development in Africa, 27(1), 23-53.

Manjengwa, J., 2012, Environment, Natural Resources and Poverty Reduction: Capturing and Sharing the Gains of Natural Resources Exploitation. In: Understanding poverty, promoting wellbeing and sustainable development A sample survey of 16 districts of Zimbabwe. Harare: Sable Press, 79-94.

Manjengwa, J. & Chimhou, A., 2012, Conclusion and Implications for Policy. In: Understanding poverty, promoting wellbeing and sustainable development A sample survey of 16 districts of Zimbabwe. Harare: Sable Press, 95-101.

Mapuva, J., 2015, Skewed rural development policies and economic malaise in Zimbabwe. African Journal of History and Culture, 7(7), 142-151.

Mashayahanya, R., 2014, More farmers take to tobacco farming. [Online] Available at: http://www.thezimmail.co.zw/2014/10/16/more-farmers-take-to-tobacco-farming/. Accessed 24 October 2014].

Matondi, P. & Rukuni, M., 2010, Rebuilding Capacity for Policy Analysis and pro poor policy making in Africa, Harare: Ruzivo Trust.

Matonho, T., 2015, REDD+ fails to transform societies. [Online] Available at: https://www.newsday.co.zw/2015/06/10/redd-projects-fail-to-trans form-communities/. [Accessed 21 May 2016].

Mbigi, L., 2014, Innovation, Culture and African Management. In: A. Nyanguru, ed. Promoting Social Work for Zimbabwe's Development. Bindura: Bindura University of Science Education, 21-34.

Mugumbate, J., & Maushe, F., 2014, Social work: fundamentals and eclecticism. In: A. Nyanguru, ed. Promoting Social Work for Zimbabwe's Development. Bindura: Bindura University of Science Education, 59-77.

Murombedzi, J., 2014, The Environment in Social Science and Humanities in Africa. [Online] Available at: http://www.codesria.org/IMG/article_PDF/article_a1970.pdf. [Accessed 24 January 2014].

Mutasa, M., 2011, Climate change vulnerability and adaptation in failing states: Zimbabwe's drought struggle. Michigan, Initiative on Climate Adaptation Research and Understanding through the Social Sciences (ICARUS-2) meeting.

Muzulu, P., 2014, Mathonsi landmark judgement: New land paradigm? [Online] Available at: https://www.newsday.co.zw/2014/03/10/ mathonsi-landmark-judgement-new-land-paradigm/. [Accessed 17 March 2014].

Pito, N., 2016, Stop aiding deforestation Firm told. [Online] Available at: www.herald.co.zw/stop-aiding-deforestation-forestry-firm-told/ [Accessed 22 May 2016].

Scoones, I., 2015, Tackling climate-change the contested politics of forest carbon projects in Africa. [Online] Available at: https://zimbabweland.wordpress.com/2015/07/27/tackling-climate-change-the-contested-politics-of-forest-carbon-projects-in-africa/. [Accessed 4 June 2016].

United Nations Country Team (UNCT), 2014, Zimbabwe country analysis working document final Draft-Information. [Online] Available at: https://ims.undg.org/.../7e40fe82fedfcf6fb92306b459a8c1bdd0d13cc9ea8e9a18cadb. [Accessed 12 May 2016].

Chapter Three

CLIMATE CHANGE COMMUNICATION: THE KENYAN STORY

Susy Wandera
Sustainable Environmental Development Watch Kenya (SUSWATCH)

Abstract

The process of developing climate legislation in Kenya began in 2008 with a dialogue between a legislator, the National Climate Change Consortium of Kenya (NCCCK) and the Kenya Climate Change Forum (KCF). The outcome of this meeting led to the merger of the two civil society forums pooling their variety of backgrounds and skills that worked with communities, youth groups, and researchers, under the auspices of the Kenya Climate Change Working Group (KCCWG). The subsequent draft Climate Change Bill was taken around the country by a team comprising the KCCWG and legislators under their caucus of the Parliamentary Network on Renewable Energy and Climate Change (PANERECC). The methodology highlighted in this paper seeks to have the participants engaged with, to get something there and in the activities undertaken by engaging targeted audiences, branding and key messages, public relations, timing and use of different media formats to involve different stakeholders. The application instrument used to engage with communities was 'climate hearings' which have been a vigorous process of communities seeking redress from time immemorial. The Government of Kenya (GoK) developed the National Climate Change Response Strategy (NCCRS) in 2009. The five-year National Climate Change Action Plan (NCCAP) was established soon after to operationalize the NCCRS and placed in the public domain in 2013. In all these processes the GoK involved a wide variety of non-state stakeholders at every stage from conceptualization to the final product. The complexity of creating awareness of the climate change (CC) discourse in Kenya has not been without hitches. However, tackling climate change is a long-term action that generations beyond the current population in Kenya's entire spectrum of society will grapple with. The legal backing of the Climate Change Act 2016 provides a wealth of opportunities to continue expanding work in this arena applying the lessons learned by the predecessors in this regard. This is but a drop in the ocean that hopefully will contribute to the literature of communicating climate information and spur the growth of such evidence from a developing country.

Key words: Communication • Hearings • Initiative • Instruments • Network

Introduction

In a 1991 address, Daisaku Ikeda proposed that 'in an age when knowledge, information, culture, ideas, and systems-elements of what had been characterised as 'soft power'-were playing an increasingly important role in shaping world history, an inner-motivated philosophy would be essential as a basis for sustaining and accelerating this trend'.[1] Further, in his annual peace proposal to the United Nations (UN) in 2002, Ikeda quoted then-President Seyed Mohammad Khatami of Iran who said: '...no great culture and no great civilization have ever evolved in isolation...' In other words, only those cultures and civilizations [that] have survived have been empowered with 'communication', 'speaking' and 'listening'. In addition to 'speaking', 'dialogue' requires 'listening'...listening is not only a passive activity; it is active, an activity which enables the listener to open his being to the world which the speaker creates or discovers. Without real listening, any dialogue is doomed to failure'. Ikeda reiterated that he likewise believed that the value of dialogue is to be found in its processes, perhaps even more than in its concrete results.[2]

This paper, therefore, highlights Kenya's effort in communicating climate change information, by specific entities, which utilised a variety of forms sensitive to the context of Kenya's cultural, social and religious dynamic. The target audiences have spanned those interested in climate science, to negotiations, to those focused on tackling adaptation and mitigation, and those who desire to take advantage of opportunities to create new products and services. A variety of forums both formal and informal was and continues to be applied to impart and gain new knowledge and information. Communicating the whole gamut of the climate change issue is a long-term venture that Kenya has come a long way in improving its delivery. Kenya's media, some who have been trained on covering climate negotiations, have helped provide space in the various channels of print, television, and radio, national and sub-regional, to educate the public on climate change matters further. Information on climate science and the ability to understand and work with

---

[1] Ikeda, D 1991, The Age of 'Soft Power' and Inner-Motivated Philosophy delivered at Harvard University

[2] Ikeda, D 2002, Peace Proposal The Humanism of the Middle Way: Dawn of a Global Civilization

uncertainty is an essential resource that assists decision making for adaptation and resilience. A multi-level, cross-sectoral approach involving a range of different stakeholders is necessary to garner success in developing adaptive capacity and build long-term resilience. In addition, keeping in mind, the two learning styles people have of the right brain 'visual' and the left-brain 'logical' and endeavouring to integrate these varied ways of absorbing information when sensitising target audiences about climate change, enhances the uptake of the message for transformative and interactive engagement. While that may seem common sense, the bumpy journey in communicating the complexity of climate change to garner buy-in and grow the critical mass of allies necessary to bring all hands on deck requires persistence.

Study Area and Methods

Kenya is reputed to have the most vibrant civil society on the African continent. This is particularly so in the climate change arena where Kenya's engagement between its non-state actors and the national and county governments is constructive and consultative. The non-state stakeholders strive to have gender parity in interaction with youth, communities, faith-based and national organizations, academia, private sector, and media. Wolf and Moser (2011) define communication as 'a personal state of connection with the issue of climate change, in contrast to engagement solely as a process of public participation in policy-making'.[3]

Kenya's climate is influenced by its geographical position relative to the Equator, its nearness to the Indian Ocean, Lake Victoria, varied topography as well as the Inter-Tropical Convergence Zone (ITCZ). Annual rainfall in Kenya follows a bimodal seasonal pattern, with long rains occurring in March to May and short rains in October to December.[4] The evidence of climate change in Kenya includes rising temperatures and shifting rainfall patterns that are leading to a growing frequency and intensity of droughts and floods and increase in incidences of diseases.[5] Of necessity, funds are diverted to handle the climate-related extreme events (droughts, floods, landslides), placing intended socio-economic development activities behind schedule. This is detrimental to the population's livelihoods and the country's aspirations for progress. Kenya's level of exposure to climate change impacts is de-

---

[3] Wolf, J and Moser 2011, S.C Individual understandings, perceptions, and engagement with climate change: insights from in-depth studies across the world

[4] Road to COP 2010, Climate Change: Campaigning and Advocacy in Kenya-Road to COP p. 8

[5] Marigi, S 2009 Key Climate Change Concerns and Issues in Kenya: Vulnerability, Impacts and Adaptation

picted in the vulnerability index (Fig. 3.1) that was generated during the development of the National Adaptation Plan (NAP).

As a signatory to the United Nations Framework Convention on Climate Change (UNFCCC) since 1994, Kenya required national policy and regulatory mechanisms to help translate its international obligations. Laozi, in the Tao Te Ching, stated that *a journey of a thousand miles begins with a single step.*[6] Kenya's journey in deliberating on incorporating its international climate change obligations into its national system can be traced to 2006. As the host of the 12th session of the Conference of the Parties to the UNFCCC and second session of the Parties to the Kyoto Protocol (COP12/CMP2) three things occurred:

Fig. 3.1: Kenya's Climate Vulnerability Index

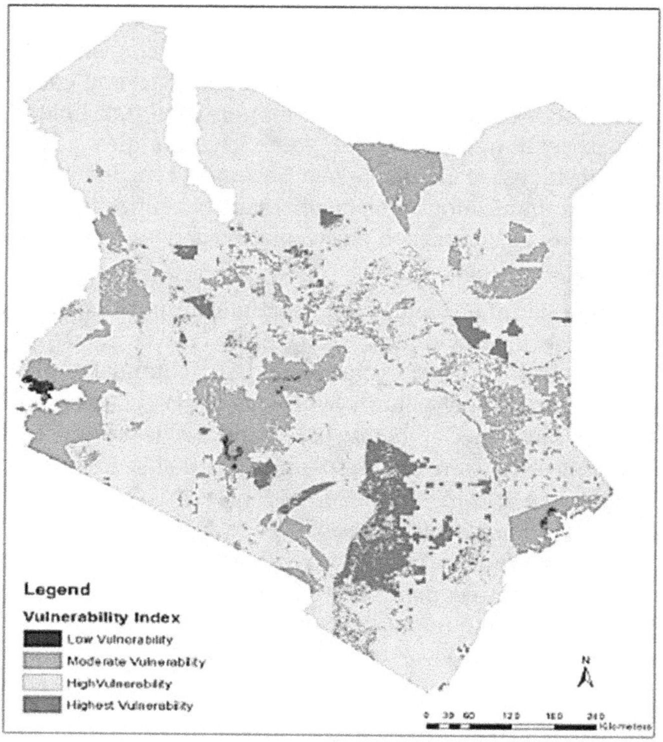

Source: Government of Kenya (GoK), Ministry of Environment and Natural Resources (MENR) 2015

---

[6] Tao Te Ching, http://taotechingme.com/?s=chapter+64

- Civil society representatives were included in the COP12/CMP2 organizing committee
- Following the conclusion of COP12/CMP2, the National Climate Change Consortium of Kenya (NCCCK) and the Kenya Climate Change Forum (KCF) were established
- The African Youth Initiative on Climate Change (AYICC) was born at the second session of the Conference of Youth (COY2), which later triggered the creation of the Kenya Youth Climate Network (KYCN)

NCCCK and KCF endeavoured to work with communities in agricultural production and those from arid and semi-arid lands (ASALs) respectively, to enable them to cope with and minimize the negative impacts of climate change. Youth groups who comprised KYCN and AYICC Kenya chapter also undertook their own activities to explore their role in the climate change discourse. The general public and many policymakers generally struggle to understand the language of scientists. And likewise, scientists often find difficulty in expressing themselves simply. It is a fact that the scientific community such as the Intergovernmental Panel on Climate Change (IPCC) is conservative and measured when communicating its interpretation of climate data and the conclusions arrived at. Some communities have experienced research fatigue from the numerous interactions with the scientific community, often complaining of feeling like guinea pigs and that results are not communicated back to them.

The methodology used in the activities highlighted in this paper fundamentally seeks to have the participants engaged with, to get something there and then in the activities undertaken. Specifically, the following steps were deployed:

- Targeted Audiences: through identification of the decision and/or opinion makers within the ecosystem of the community
- Branding and Key Messages: messaging to link why and how tackling climate change is important and that everybody has a role
- Public Relations: good relations with the 'front line' of the decision and/or opinion makers to create a conducive space for delivery and receptivity of the message by the audience
- Timing: interest in climate change issues becomes heightened during periods of extreme weather events
- Use of Different Media Formats: publicise the activity and/or being interviewed

- Champions: inclusion of high visibility personalities to embed the message
- Stakeholder Engagement: decision and/or opinion makers and communities create networks through establishing contacts

People learning together to communicate climate change in a way that will enhance resilience and achieve the transformation of behaviour takes time, and requires the building and nurturing of trust as well as managing conflict arising from vested interests.

Coordinating Climate Change Activities

The process of developing climate legislation in Kenya began in 2008 with a conversation between a legislator, the National Climate Change Consortium of Kenya (NCCCK) and the Kenya Climate Change Forum (KCF). The aftermath of this dialogue led to the merger of the two civil society forums pooling their combined variety of backgrounds and skill sets that worked with communities, youth at the national level and researchers, under the auspices of the Kenya Climate Change Working Group (KCCWG). This foundation of coordinated climate change activities within the civil society was and continues to be a critical contribution to enhancing Kenya's ability in tackling climate change. It has attracted positive support from the national and county governments, as well as development partners.

Faith-based, humanitarian, research and academic organizations such as Norwegian Church Aid (NCA), Catholic Agency for Overseas Development (CAFOD), TROCAIRE, Heinrich Böll Stiftung (HBS), Oxfam GB, Act Change Transform (ACT!), Practical Action, Catholic Organization for Relief and Development Aid (CORDAID), African Centre for Technology Studies (ACTS) and the University of Nairobi's Institute for Climate Change and Adaptation (ICCA) were, and, are instrumental in these collaborative actions to build the capacity of their partners and constituents.

KCCWG in responding to the aforementioned legislator's concerns undertook desk research to determine the state of play in Kenya on existing policy instruments that contained guidance on grappling with climate change across the socio-economic sectors. The findings fed into the development of draft legislation, and its design, that was then handed over to the legislators to fine-tune into the required legislative format and be submitted as a Private Member's Bill. At that time, Bills or legislation were usually generated from the executive branch of government and then tabled in Parliament. In this case, it was a legislator who tabled the motion to generate such law as a Private Member's [Member of Parliament] Bill. The record of Kenya's verbatim parliamentary proceedings–the Hansard on this subject in late 2008 still exists. This also disputed the claim by opposing elements in the Executive that it was a 'Civil Society Bill' to justify their antagonism throughout the process the first time. The youth, a crucial constituency, was continually involved in the effort to understand the climate legislation, its process and intention, and its connection to their activities. Youth were encouraged to share their experiences with their peers within the KYCN and AYICC family. An increasing number of other non-state actors (NSAs) continue to emerge, and they are striving to incorporate climate change into their core activities and/or exploring variety issues within the climate change agenda.

Climate Hearings

The subsequent draft Climate Change Bill was taken around the country by a team comprising the Kenya Climate Change Working Group (KCCWG) and the legislators under their caucus of the Parliamentary Network on Renewable Energy and Climate Change (PANERECC). The application tool used to engage with communities was that of 'climate hearings'. Hearings have been an integral process of communities seeking redress from time immemorial. The Panchayats in South Asia and Elder's Councils in Africa are the true precursors to modern day hearings.[7] These climate hearings provided the space to:

- Explain climate science in a simple way
- Break the myth of the divine hand meting punishment
- Encourage communities to express how climate change impacts affected them and recognise their contributions toward tackling climate change

While communities may be unfamiliar with climate change terms, they do know what climate change is, and this was evident through all the regions visited. Figs. 3.2, 3.3 and 3.4 are examples of community engagement. Other stakeholders reached out to were faith-based, humanitarian and the wider non-state actors' constituencies, as well as private sector in all the regions the hearings, took place. Representatives of the aforementioned were included in the local organizing committees to collaborate with the Nairobi team. This work continued over a period of about five years from 2009 until 2014. The first phase was from 2009–2012, the second from 2013-2014.

[7] Otiende, B, 2009, Climate Change Campaign in Kenya: Climate Hearings

Fig. 3.2: Community Climate Hearing in Tharaka Nithi

Source: Susy Wandera (2014)

Fig. 3.3: Community Climate Hearing in Ishiara

Source: Susy Wandera (2014)

Fig. 3.4: Community Intervention at East Pokot

Source: Susy Wandera (2015)

Arid and Semi-Arid Lands (ASALs)

Specific actions undertaken in this region by KCCWG included:

- conducting research on access and use of climate change (CC) information
- documenting communities' best practices in adaptation and mitigation and their replicability
- carrying out climate hearings[8]

Rauka ama Hatutasurvive

KCCWG's Road to COP (Conference of Parties) whose members included KYCN and AYICC held a heart storming session to create a clarion call that encapsulated the following:

---

[8] Kenya Climate Change Working Group (KCCWG) 2013, Report on Climate Change Actors and Opportunities for Scaling up County Specific Climate Change Mitigation and Adaptation Measures in Selected ASALs Counties in Kenya

- our survival, lives, and livelihoods were at stake
- we did not want to be locked into poverty
- we had the right to develop
- developed countries were required to provide our country with financial resources to facilitate our adaptation [historical responsibility], and they needed to cut their domestic emissions in their own geographical space
- as a priority, our leaders needed to provide and operationalize responsible governance in resource management

The words that emerged to symbolise these sentiments were *'Rauka ama Hatutasurvive!'* – Kenyanese for *Wake Up or We Won't Survive* and continue to underpin implicitly and explicitly all activities conducted. Rauka ama Hatutasurvive spoke to two basic tenets:

- responding to the impacts of climate change was a collective endeavour that cuts across the background, education, income, age, gender, culture and any other grouping
- generationally, youth will be living in the four or five decades in the future and will be directly affected, so active involvement now creates room for influencing change in behaviour, attitude, etc.[9]

Kenya Youth Climate Network (KYCN)

The groups of youth that comprised KYCN held a session of deep reflection on a guiding principle to enrich their climate change engagement. This principle became known as the *Sagana Resolution on Youth Leadership in Kenya* that exhorts the creation of alternative leadership committed to being green in everything that is done and said, embedded into all spiritual, verbal, written and physical actions.[10] KYCN undertook five national youth conferences on climate change (NYCCCs) from 2008 until 2012 and several regional youth conferences on climate change (RYCCCs) from 2013 to 2014, bringing together national and community youth organizations, students groups, and individuals interested in working on climate change matters. These deliberations have set a solid foundation for growing a critical mass of 'climate change allies'.

---

[9] Road to COP 2010, Climate Change: Campaigning and Advocacy in Kenya-Road to COP p. 14

[10] Kenya Youth Climate Network (KYCN), 2010 Sagana Resolution

African Youth Initiative on Climate Change (AYICC) Kenya Chapter

AYICC-K mobilises university and youth groups in a variety of activities. In 2016, they hosted the second African Youth Conference on Climate Change (AfriYOCC2) in Nairobi. 135 young people attended this meeting from 21 African countries. A key product from these African youth working under and within the platforms of AYICC and the Committee of African Heads of State and Government on Climate Change's (CAHOSCC) Youth Programme on Climate Change (CYPCC), was the African Youth Position on the 22nd session of the UNFCCC Conference of the Parties (COP22) to the African Group of Negotiators (AGN) urging the inclusion of youth's concerns in the stance AGN takes during the various sessions.[11]

COP17 African Youth Climate Justice Campaign

Both KYCN and AYICC-K's campaign activities in various parts of the country, at the regional and global level, culminated in a major endeavour toward COP17 entitled *'We Have Faith! Act Now for Climate Justice'* a carbon neutral road journey of six trucks from Nairobi to Durban comprising 160 participants from 18 countries. Fig. 3.5 is the certification of carbon neutrality. KYCN and AYICC-K, in addition to using formal formats, as young people also take advantage of the leverage gained from social media and other mediums, as they continue to seek space to engage all stakeholders in national, regional and international activities at the preparatory and meeting level, that inform the climate change discourse such as Rio+20, UNFCCC COPs, African Union Commission (AUC) meetings and many others.

The hip-hop artiste Juliani (Fig.3.6), a KYCN member travelled with the caravan to Durban where he performed at concerts along the route of the 6 countries traversed. The late Achieng' Abura (Fig. 3.7), an Afro-jazz artiste and environmental advocate performed at the Nairobi concert on the eve of the caravan's departure. Both these personalities not only lent their voice they supported the campaign in a myriad of ways behind the scenes. Fig. 3.8 shows the symbolic flagging off of the COP 17 Caravan, an expression of overcoming the tremendous challenges experienced–financial and from detractors–and succeeding in undertaking this climate activity.

[11] African Youth Initiative on Climate Change (AYICC), 2016 African Youth Position on COP22

Progress of Climate Change Instruments

The Government of Kenya (GoK) developed the National Climate Change Response Strategy (NCCRS) in 2009 and launched it in 2010. The five-year National Climate Change Action Plan (NCCAP) was established soon after to operationalize the NCCRS and placed in the public domain in 2013. In all these processes the GoK involved a wide variety of non-state stakeholders at every stage from conceptualization to the final product.

Fig. 3.5: Certificate of Carbon Neutral Footprint

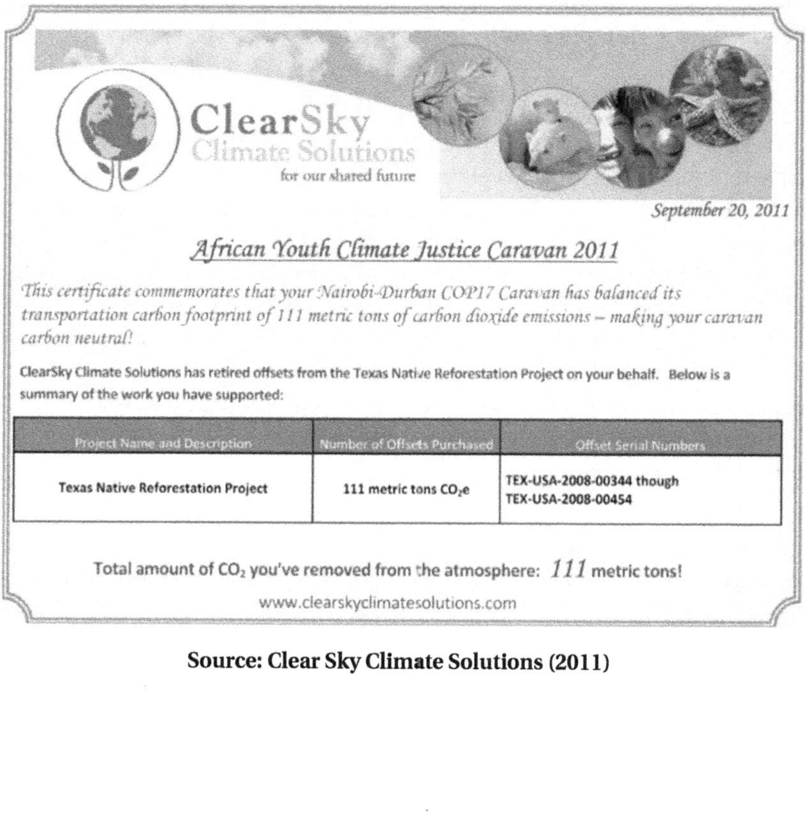

ClearSky
Climate Solutions
for our shared future

September 20, 2011

African Youth Climate Justice Caravan 2011

This certificate commemorates that your Nairobi-Durban COP17 Caravan has balanced its transportation carbon footprint of 111 metric tons of carbon dioxide emissions – making your caravan carbon neutral!

ClearSky Climate Solutions has retired offsets from the Texas Native Reforestation Project on your behalf. Below is a summary of the work you have supported:

Project Name and Description	Number of Offsets Purchased	Offset Serial Numbers
Texas Native Reforestation Project	111 metric tons CO_2e	TEX-USA-2008-00344 though TEX-USA-2008-00454

Total amount of CO_2 you've removed from the atmosphere: *111* metric tons!

www.clearskyclimatesolutions.com

Source: Clear Sky Climate Solutions (2011)

Fig. 3.6: Juliani at COP 17 Nairobi Concert

Source: Jack Stack (2011)

Fig. 3.7: The late Achieng' Abura at COP 17 Kick-off Nairobi Concert

Source: Jack Stack (2011)

Fig. 3.8: Symbolic Flag off of COP 17 Caravan

Source: Jack Stack (2011)

Before and After 2013 Elections

The challenge of then President Kibaki's refusal to assent to the Climate Change Bill in early 2013 citing insufficient consultation meant going back to the drawing board. While the then Head of State and Government (HS & G) did not specify reasons for his conclusion, as indicated in a previous section, the legislative process had faced adverse responses from some in the Executive. It was later determined that some private sector entities had claimed exclusion and opposed the Bill, which the facts also dispute. This sector was part of local organizing committees of the grassroots engagements. Prior to Kenya going to the elections, the Kenya Climate Change Working Group (KCCWG) reached out to a variety of stakeholders in the civil society, private sector, legislature and government in mid-2013 to deliberate on the lessons from the previous work on sensitising Kenyans on the climate legislation. Fig. 3.9 shows the Mover of the Climate Change Bill, Hon. Dr. Wilbur Ottichilo giving remarks during a session after elections on determining the next step for the climate law process.

Following the elections, which operationalized the new constitutional dispensation of devolved governance, KCCWG organized further engagements on re-examining the terms of reference (TORs) that would revitalise work on the Climate Change Bill. An additional plus was that the incoming Executive branch was willing to collaborate with both the civil society and the legislature. The result was that the national focal point to the UNFCCC, the Cabinet Secretary (CS) for Ministry of Environment and Natural Resources (MENR), Prof. Judi Wakhungu constituted a gazetted taskforce to produce a new draft of the Climate Change Policy and Bill.

Fig. 3.9: Hon. Dr. Wilbur Ottichilo, the Mover of the CC Bill

Source: Susy Wandera (2013)

In Kenya, a gazetted team is an official entity of appointed individuals to carry out specific assignments. This CC Policy and Bill team comprised people from ministries and state agencies, a legislator, private sector and civil society. In hindsight, the refusal to assent was a blessing in disguise as it enabled the integration of the devolved governance structure and expansion of more stakeholders' involvement. Figs. 3.10 and 3.11 show representatives from the Council of Governors' (CoG) Committee of Executive Committee Members (CECs) responsible for the environment from all the 47 counties who had their interventions included in the draft legislation. This meeting hosted by MENR had its Cabinet Secretary Prof. Wakhungu, her Principal Secretary (PS), a Member of Parliament from the National Assembly's Environment Committee and the CC Bill and Policy taskforce in attendance.

As before, youth were involved every step of the way, from providing draft text on the policy that informed the Bill to creating awareness. Youth are a diverse group, some interested in hands-on work at the grassroots level, others on innovative technologies, others on policy, others on research, others on advocacy campaigns and still others on the international negotiations. The dedicated effort was and continues to be expended to enhance the capacity to support and ensure that this youth access whatever resources they need to take action in their preferred areas of interest successfully. The emphasis on youth is simply because we need a critical mass of them to embrace the cause of tackling climate change as it is their generation and the generations after that can carry on the myriad of activities required for this work to have meaning for the next 50, 100 and 200 years.

Fig. 3.10: CS, PS, CECs, MP and CC Bill and Policy Taskforce

Source: GoK, MENR (2014)

Fig. 3.11: CC Bill and Policy Breakout Session

Source: GoK, MENR (2014)

May 6, will always be a significant date as that was when President Uhuru Kenyatta assented to the Climate Change Act 2016. The essence of the Climate Change law is that it is people-centered, and it expects all stakeholders to fulfil their role in helping Kenya build its ability to handle climate change and its transition into a low carbon development path. This piece of legislation which is being integrated into all of Kenya's socio-economic tools is, without fear of contradiction, the most consultative, scrutinised and people-centred instrument in the world to date (2016). The clauses throughout this Climate Change law attest to this as highlighted by the following examples:

- *'This Act shall be applied to the development, management, implementation and regulation of mechanisms to enhance climate change resilience and low carbon development for the sustainable development of Kenya'.*[12]
- *'...this Act shall be applied in all sectors of the economy by the national and county governments to–formulate programmes and plans to enhance the resilience and adaptive capacity of human and ecological systems to the impacts of climate change*[13]*; mainstream intergenerational and gender equity in all aspects of climate change responses...'*[14]
- *'The guiding values and principles.... shall bind all level of government and all persons...'*
- *'In discharging their functions and duties...any state officer and state organ shall be guided by the following: ensure participation and consultation with stakeholders....'*[15]
- *'There is established...the National Climate Change Council*
- *The Council shall be chaired by the President [of the Republic of Kenya]'*[16]
- *'The Kenya Institute of Curriculum Development shall...integrate climate change into various disciplines and subjects of the nation The last clause is of particular significance as Kenya is currently reviewing its entire education syllabus. The timing could not be better for the collaborative work on-going between the Climate Change Directorate (CCD) a body created by this law and under the auspicious of the MENR the UNFCCC national focal point, the Kenya Institute of Curriculum Development (KICD), the University of Nairobi's Institute of Climate Change and Adaptation (ICCA) and the Council of Governors (CoG) on one hand, and the numerous variety of stakeholders in the Education sector on the other.*
- *As Kenya made its preparations for the 22nd Session of the Conference of the Parties (COP22) of the UNFCCC in Marrakech, the focal point MENR's CCD held activities with a wide array of stake-*

[12] Republic of Kenya, The Climate Change Act 2016, Objects and purposes Article 3 paragraph 1, p. 183

[13] Ibid.

[14] Ibid

[15] Ibid, p. 185

[16] Ibid

holders in this regard. In collaboration with the National Gender and Equality Commission (NGEC) a constitutional body, there was an engagement to submit views to the 45th session of the Subsidiary Body on Implementation (SBI45) on enhancing the Lima Work Programme (LWP). NGEC which heads the secretariat for the African Working Group on Gender and Climate Change (AWGGCC) also held a session with the African Group of Negotiators (AGN) on LWP.

- *Opportunities emerge to convey climate change information in deliberations on sustainable land management (SLM), in particular demonstrating the synergy between the UNFCCC and the United Nations Convention to Combat Desertification (UNCCD). This interaction is an on-going collaboration between the Sustainable Environmental Development Watch (SUSWATCH) Kenya, the Kenya Agricultural Productivity and Sustainable Land Management Project (KAPSLMP) of the MENR, and the State Department of Livestock. Linkages between climate change and sustainable land management are expressed in two Rio instruments as follows:*
- *al education curricula at all levels...'*[17]

The last clause is of particular significance as Kenya is currently reviewing its entire education syllabus. The timing could not be better for the collaborative work on-going between the Climate Change Directorate (CCD) a body created by this law and under the auspicious of the MENR the UNFCCC national focal point, the Kenya Institute of Curriculum Development (KICD), the University of Nairobi's Institute of Climate Change and Adaptation (ICCA) and the Council of Governors (CoG) on one hand, and the numerous variety of stakeholders in the Education sector on the other.

As Kenya made its preparations for the 22nd Session of the Conference of the Parties (COP22) of the UNFCCC in Marrakech, the focal point MENR's CCD held activities with a wide array of stakeholders in this regard. In collaboration with the National Gender and Equality Commission (NGEC) a constitutional body, there was an engagement to submit views to the 45th session of the Subsidiary Body on Implementation (SBI45) on enhancing the Lima Work Programme (LWP). NGEC which heads the secretariat for the African Working Group on Gender and Climate Change (AWGGCC) also held a session with the African Group of Negotiators (AGN) on LWP.

---

[17] Ibid, p 198

Opportunities emerge to convey climate change information in deliberations on sustainable land management (SLM), in particular demonstrating the synergy between the UNFCCC and the United Nations Convention to Combat Desertification (UNCCD). This interaction is an on-going collaboration between the Sustainable Environmental Development Watch (SUSWATCH) Kenya, the Kenya Agricultural Productivity and Sustainable Land Management Project (KAPSLMP) of the MENR, and the State Department of Livestock. Linkages between climate change and sustainable land management are expressed in two Rio instruments as follows:

- *Article 2 of UNFCCC '...stabilise atmospheric concentrations of greenhouse gas emissions (GHGs) to avoid a dangerous anthropogenic interference with the climate system'*
- *Article 2 of UNCCD '...involve long-term integrated strategies that focus simultaneously on improved productivity of land, rehabilitation, conservation and sustainable management of land and water resources, leading to improved living conditions, in particular at the community level'*
- *Article 2 of the Paris Agreement '...enhancing the implementation of the Convention...to strengthen the global response to the threat of climate change, in the context of sustainable development and efforts to eradicate poverty by increasing the ability to adapt to adverse impacts of climate change and foster climate resilience and low greenhouse gas emissions development, in a manner that does not threaten food production'*

Fig. 3.12: Western Region County Governments and Non-State Actors SLM Meeting

Source: GoK MENR (2016)

Limitations to Climate Change Communication

The complexity of creating awareness of the climate change (CC) discourse in Kenya has not occurred without hindrance. The challenges experienced were:

- vested interests who sought to impede the progress of the climate change legislative process on the one hand, and the youth's COP17 campaign on the other
- gatekeeping
- garnering comprehension and agreement from target audiences that climate change was a development issue not just an environmental problem
- translating the science into simple correct terms, for instance, in Kiswahili CC is *'Mabadiliko Tabi ya Nchi'*
- first reactions that CC was the development partners' agenda, not Kenya's
- early on there was indifference from some stakeholders who could not see how CC related to their activities
- initial lacklustre coverage from media since CC was not 'sexy'
- sufficient finance to reach wider audiences and provide easily understood material

Conclusion and the Way Forward

Kenya through its various formal and informal engagements from its academic, civil society, youth, private and public sectors has utilised all opportunities to exchange and share climate change information. The work highlighted in this chapter are snapshots of the kind of effort made thus far in doing so to a growing audience whom it is hoped then go on to share the insights they glean with their own ecosystem of constituents and/or communities. Because of the overwhelming body of work that continues to emerge, all the aforementioned avenues will be useful in Kenya's endeavour to have a population that is able to build resilience while integrating low carbon development mechanisms effectively.

Henry Wadsworth Longfellow wrote: *'...not enjoyment, and not sorrow, is our destined end or way; but to act, that each tomorrow finds us further than today...'*[18] Tackling climate change is a long-term action that generations beyond the current population in Kenya's entire spectrum of society will grapple with. The legal backing of the Climate Change Act 2016 provides a wealth of

[18] Wadsworth, H.L 1966, The Poetical Works

opportunities to continue expanding work in this arena applying the lessons learned by the predecessors in this regard. This chapter is but a drop in the ocean that will hopefully contribute to the literature of communicating climate information and spur the growth of such evidence from a developing country.

References

African Youth Initiative on Climate Change (AYICC) 2016, African Youth Position on COP22, Nairobi.

Arid Lands Information Network (ALIN) & CARE International Adaptation Learning Programme 2013, Dissemination of climate information using radio in Kenya, Joto Afrika, 8, Nairobi.

CARE International 2015, Adaptation Learning Programme for Africa ALP Results, Outcomes and Impacts Report January 2010 to June 2015, Geneva.

Ching T, T., Addiss, L.W & Tao, T.C., 2016, http://taotechingme.com/?s= chapter+64.

Government of Kenya, Ministry of Environment and Natural Resources 2016, Kenya National Adaptation Plan (2015-2030), Enhanced climate resilience towards the attainment of Vision 2030 and beyond.

Government of Kenya, Ministry of Environment and Natural Resources 2013, Kenya Climate Change Action Plan (2013-2017), http://www.kccap. info, Nairobi.

Government of Kenya, Ministry of Planning 2013, Mainstreaming climate change resilience into development planning in Kenya, Nairobi.

Heinrich Böll Stiftung East & Horn of Africa, 2013, Towards a Coherent and Cost-effective Policy Response to Climate Change in Kenya: Country report, Nairobi.

Ikeda, D. 1991, The Age of 'Soft Power' and Inner-Motivated Philosophy delivered at Harvard University, Cambridge, Massachusetts, USA, September 26, 1991, http://www.daisakuikeda.org/sub/resources /works/lect/lect-01.html, Tokyo.

Ikeda, D. 2002, Peace Proposal the Humanism of the Middle Way: Dawn of a Global Civilization Soka Gakkai International (SGI), Tokyo.

Kenya Climate Change Working Group (KCCWG) 2014, Training Manual to Facilitate County Governments' Response to Climate Change Mitigation and Adaptation Exigencies in ASALs Counties in Kenya, Nairobi.

Kenya Climate Change Working Group (KCCWG) 2013, Report on Climate Change Actors and Opportunities for Scaling up County Specific Climate Change Mitigation and Adaptation Measures in Selected ASALs Counties in Kenya, Nairobi.

Kenya Climate Change Working Group (KCCWG) 2013, Fact Sheet: Access and Use of Climate Change Information in the ASALs of Kenya, Nairobi.

Kenya Climate Change Working Group (KCCWG) 2013, Gender and Climate Change Thematic Area Policy Briefs, Nairobi.

Kenya Climate Change Working Group (KCCWG) 2013, Report on ASALs Climate Change Actors Mapping, Nairobi.

Kenya Youth Climate Network (KYCN), 2010 Sagana Resolution, http://kenyayouthclimatenetwork.blogspot.co.ke/2010/12/s0010042010-sagana-resolution-onyouth.html, Malta Camp Sagana.

Kenya Private Sector Alliance (KEPSA) 2014, Climate Change and the Kenyan Private Sector Climate Change and Your Business Briefing Note Series, Nairobi.

Marigi, S. 2009 Key Climate Change Concerns and Issues in Kenya: Vulnerability, Impacts and Adaptation, Nairobi.

Norwegian Church Aid 2010, Kenya Youth Climate Change Handbook, Nairobi.

Otiende, B. 2009, Climate Change Campaign in Kenya: Climate Hearings Oxfam GB, Nairobi.

Oxfam, G.B. 2009, Oxfam and Climate Change: Approach and Priorities, Nairobi.

Parliament of Kenya 2016, Kenya Climate Change Act, http://kenyalaw.org/kl/fileadmin/pdfdownloads/Acts/ClimateChangeActNo11of2016.pdf, Nairobi.

Road to COP 2010, Climate Change: Campaigning and Advocacy in Kenya-Road to COP, Nairobi.

Stack, J.N. 2012, Road to Durban Act Now for Climate Justice. COP17 Campaign 2011 in Pursuit for Climate Justice, https://www.youtube.com/watch? v=MxFMAypNpc8, Nairobi.

Wadsworth, H.L. 1966, The Poetical Works. AMS Press, Inc. New York, (1), 21.

Wandera, S. 2015-2016, Climate Change and Sustainable Land Management (SLM): Challenges, Opportunities and Partnerships, Nairobi.

Wandera, S. 2015, Youth Involvement in Climate Change Negotiations, Youth Conference on Ninth Session of the Community Based Adaptation to Climate Change (CBA9), Nairobi.

Wandera, S. 2014, 'Walking with all: A journey towards climate change legislation and policy in Kenya' in ACT! Adapting to Changing Times: Models for climate change adaptation in Kenya, 41-45, Nairobi.

Wandera, S. 2013, Why Involve the Youth in Climate Change, http://www.theeastafrican.co.ke/OpEd/comment/Why-involve-the- youth-in-climate-change-/-/434750/2085322/-/sm3lpqz/-/index.html, Nairobi.

Wolf, J. & Moser, S.C. 2011, S.C Individual understandings, perceptions, and engagement with climate change: insights from in-depth studies across the world, WIREsClim Change2011 DOI: 10.1002/wcc, 120, Hoboken.

Chapter Four

IT TAKES TWO TO TANGO: POTENTIAL AND ACTUAL USE OF WEATHER AND CLIMATE INFORMATION FOR DECISION-MAKING IN ZIMBABWE

Vincent Itai Tanyanyiwa,
Zimbabwe Open University, Zimbabwe

Rejoice Madobi,
Zimbabwe Open University, Zimbabwe

Abstract

Climate change is real, and its worsening impacts have been observed to affect most the rural poor and the vulnerable. Some support for adaptation is relevant, and this is where climate change communication becomes imperative. For adaptation to bear fruit, behaviour change is required. Behavioural change theories endeavour to clarify why behaviours change. Behavioural determination is influenced by environmental, personal, and behavioural characteristics. Currently, there has been increased interest in the application of these theories in the areas of climate change, education and health. Climate change communication is increasingly becoming topical because of the impact of climate change in science and society. This study looks at how potential and actual weather and climate information is situated in the farmers' daily lives. An analysis of the importance and complications in-built in the discussion of climate change to different stakeholders especially farmers by using various types of communication tools and approaches. This study expounds the problematic issue of the connection between climate change communication and behaviour change. In particular, it focuses on the role of language in conveying climate change information to stakeholders. As a result, this study attempts to offer an outline of developing theories of climate change communication. An evaluation of how communication could be improved in light of the theories and practices discussed concludes the paper.

Key words: Behavioural Change • Climate Change • Climate Change Communication • Livelihoods • Vulnerability

Introduction

More than 70% of Africa's livelihoods depend on subsistence agriculture for food, jobs, and income (Yeboah, 2017). Africa can feed itself, and the world for it is home to over 60% of the world's uncultivated arable land. However, food production is desperately low in the region largely because of poorly developed farming technologies which drive rain-fed farming practices. There are poorly developed climate and weather alert systems to help farmers plan for crop seasons and adopt better ways of farming, i.e., a lack of access to reliable weather data. Climate change impacts are already being felt across the globe and according to Madhavan and Rengalakshmi (nd) the outcomes of climate change include shifting rainfall patterns, temperature regimes changes and increasing climate variability. Developing countries' including Zimbabwe whose economies are dependent on climate-sensitive resources, especially rain-fed agriculture, are the most affected, and with that reality, the need to adapt has become more apparent. Crop failure due to floods and prolonged dry spells (during the rainy season) has become common in many Southern African countries and Zimbabwe in particular. Meza *et al.,* (2008) point out that climate variability imposes costs on farmers, predominantly through information constraints in production decisions and the burden that uncertainty imposes on risk-averse optimizers.

The two challenges that are faced by humankind are 'preparing for the increasingly foreseeable weather and climate, and also modifying decision-making processes in order to incorporate the availability of new information about the future' (Suarez, 2009), if the parties are to fulfil their commitments under the UNFCCC. Article 6 (of the UNFCCC) focuses on the promotion of education, training and public awareness on climate change. The article has six priority areas, and one of the areas include public access to information on climate change and its effects (UN, 1992), thus seasonal climate forecasts (rainfall and temperature) are necessary. Climate and weather information is necessary if agriculture and other sectors are to manage risks and adapt to climate change effectively. Information can increase awareness of climate risks and recognition of the need to adapt. It can also inform risk management decisions about when and how to adapt best and allow adaptation actions to be prioritized (Productivity Commission, 2012).

Knowledge of seasonal climate variability prior to crop season and its incorporation in management decisions across the value chain is touted as the key adaptation strategy for intra-seasonal climate variability (Madhavan and Rengalakshmi, n.d). But though adaptation has been reiterated at different forums, linking climate science information and farmers' practices remain a challenge. In most cases, more focus has been put on the production of accurate weather and climate information at the expense of how that information is conveyed to the farmers and how the farmers use it. The main objective of this chapter was therefore to examine the use (potential and actual) of weather and climate information to the Zimbabwean farmers. The questions which guided this study were:

a) What are the main sources of weather and climate information for the farmers in Zimbabwe?
b) What type of information is disseminated and is the information reliable?
c) How the information is disseminated and is it accessible to the farmers?
d) To what extent are the farmers making use of the information they receive?
e) What are some of the challenges that are being faced in the communication of weather and climate information?

The Climate Change Context in Zimbabwe

Climate change is one of the global threats facing humankind today, but despite its global nature, its impacts are local. Zimbabwe has one of the highest degrees of rainfall variability in the world, and the climate is characterised by frequent drought years and occasional flood events (UNICEF and IES, 2014; Brazier, 2015). This situation will be exacerbated by climate change. Records show that temperatures have been increasing by around 0.1°C a decade (Government of Zimbabwe, 2013a). Future scenarios predict increasing temperatures of around 2.5°C by 2050.

The rainy season in Zimbabwe is usually between mid-November to April. The timing and amount of rainfall received in Zimbabwe is becoming increasingly uncertain, and according to UNICEF and IES (2014), there has been an overall decline of nearly 5 percent in rainfall across the country since 1901 with the early 1990s probably witnessing the driest period of the last century. The frequency and length of dry spells during the rainy season have increased while the frequency of rain days has been reducing. Although the quantity of rainfall in a year may be normal, this may fall during violent storms. Rainfall distribution is erratic both in time and space across all the provinces of Zimbabwe. The erratic nature of rainfall patterns has a serious adverse impact on livelihoods most of which are dependent on sensitive climate resources. Climate variability and change, therefore, increase the vulnerability of poor households in both urban and rural areas.

Providers of weather and climate information

The supply of climate information consists of various actors responsible for the collection of weather data, analysis, and generation of forecasts, to dissemination and use. In many countries, particularly developing ones', systematic observations of the weather and climate system are usually the responsibility of national meteorological centres. In Zimbabwe, the Meteorological Services Department 'is mandated to provide up-to-date information on climate change issues' (Government of Zimbabwe, n. d). According to the Government of Zimbabwe (n.d), it is also the responsibility of the Zimbabwe Meteorological Services Department to provide information on Zimbabwe's emissions of greenhouse gases (GHGs) and information on interventions required to react to the adverse impacts of climate change. In most Asian countries also, the meteorological or hydro-meteorological department of the government is responsible for collecting, processing and supplying data as well as maintaining the infrastructure of the systematic observation system, UNFCCC (2007). However, other institutions such as non-governmental organisation (NGOs), research centres and agricultural extension services also play a critical role.

Seasonal Weather Forecasting in Zimbabwe

Weather information is commonly given as seasonal forecasts. This refers to the total amount of rainfall expected during the season or over a given time period (Ziervogel and Calder, 2003). Inter-annual variability in southern Africa is largely influenced by the El Nino Southern Oscillation (ENSO) phenomenon. ENSO denotes a coupled atmospheric system that links change in atmospheric pressure and sea temperature over the southern Pacific Ocean. El Nino is the widespread warming of the upper ocean in tropical Pacific. This results in increasing rainfall in the eastern Pacific and decreasing rainfall over the western Pacific and Southern Africa. In Southern Africa, most rain is brought by the Inter-Tropical Convergence Zone (ITCZ). The ITCZ is an expanse characterized by much convection activity resulting in rainfall during summer months. The likelihood of specific weather regimes is affected by sea surface temperatures, ice cover and land-sea characteristics. These boundary conditions mostly evolve slowly so they can be used to provide a forecast of different weather regime probabilities (Mason *et al.*, 1996). Reliability and the potential use of seasonal forecasts differ according to geographical region, i.e., according to oceanic atmospheric circulation, and according to the political and socio-economic context that shapes its diffusion and application. Weather forecasts are issued in September based on three categories that are below normal (BN) is when an area receives below 75% of its long-term average for the sub-season, near normal (N) is when an area receives between 75% and 125% of its long-term average in a sub-season (OND or JFM). And above normal (AN) is when an area receives more than 125% of its long-term mean for a particular sub-season (MSD, 2012). These forecasts give probable information on the forthcoming climate of time scales of 3 to 6 months. They cover the first and second half of the rainfall season, i.e., October, November, December (OND) and January, February, March (JFM) respectively (Manatsa *et al.*, 2012).

Climate predictions are issued yearly through the Southern African Regional Climate Outlook Forum (SARCOF) process. SARCOF is a regional, seasonal weather outlook prediction and application practice adopted by the Southern African Development Community (SADC), a 15-member grouping of countries in central and southern Africa. SARCOF produces an agreed forecast for the sub-region.

Importance of weather forecasting

Historical, current and future weather and climate information are necessary for climate change projections and modelling. Seasonal weather forecast and climate information (the beginning and end of each year's rainfall season have increasingly become more important in Zimbabwe as the climate scenario is increasingly becoming more and more unpredictable. This information according to World Meteorological Organization (2010) translates into:

- adequacy of rainfall amounts and deficits and excesses,
- early warnings of potentially poor seasons or disasters for example floods and droughts.

The information is necessary for the farmers' decision making process that is, 'to inform key actions for general planning questions, such as when to start planting, knowing how much to diversify, knowing which crops to plant, and the likelihood of meeting or failing to meet quotas 'World Meteorological Organization (2010).

Accurate weather forecasts and early warning increase farmer preparedness and leads to better social, economic and environmental outcomes within agricultural production systems. Important farm level decisions are based on the forecasts, and these decisions include a choice of cropping pattern, investment in fertilizers and pesticides, choice of the period for planting and plant population density. It is important for the forecasts (which are probabilistic in nature) to be well understood by the farmers if they are to be fully incorporated in farming decisions. Seasonal forecasts, therefore, give farmers an opportunity to adopt the improved technology, intensify production, replenish soil nutrients and invest in more profitable enterprises when climatic conditions are favourable and to more effectively protect their families and farms against the long-term consequences of adverse extremes. This is quite critical for Zimbabwe because crop production is predominantly rain-fed and seasonal rainfall is highly variable. Advance warning systems of extreme rainfall events should be given priority especially to the vulnerable farmers.

Limitations of seasonal climate forecasts

Seasonal climate forecasts are generally issued for fairly huge similar rainfall regions usually over 9,000km^2). The forecasts extend over a 3-month period. Hence, demarcation becomes problematic for users located on the borderline of the forecast's spatial coverage since the forecast changes each time drastically (Githungoet et al., 2009). There is usually a mismatch between farmers' needs and the scale, content, format, or accuracy of available information products and services. These factors have limited the widespread use of seasonal forecasts among smallholder farmers (Vermeulen *et al.*, 2010). In a study carried out by Mupakati and Tanyanyiwa (2017) in Chilonga Ward in Chiredzi District, Zimbabwe farmers were very uncertain of the value of these forecasts hence their low rate of utilization. The forecast skills are not only very low, but also heavily biased towards normal conditions with high confidence of the forecast is persistently in the near normal range, but this confidence is never attributed to drought forecasts resulting in above normal rainfall events being rarely forecasted and droughts never in the forecast. Manatsa *et al.*, (2012), also noted that climate variability over short distances intrinsic in the Chiredzi area reduces the usefulness of forecast information, especially when provided at the current, national scale. Seasonal length is ambiguously implied in the forecast. This explains the absence of the definition of other intra-seasonal variations in the forecast information, such as the distribution of both wet and dry spells, including their spatial distribution. Farmers and extension officers are not able to use seasonal climate forecasts given in probabilities and do not know how to respond to them (IDRC, 2010). Forecast precision depends on lead times used in the prediction. The longer lead times renders the forecast less accurate. Generally, in Southern Africa, there is a significant gap between the information needed by small-scale farmers and that provided by the meteorological services (Blench, 1999), (Fig. 13). Forecasts are generally grouped into three groups, that is, those:

a) Forecasting the next few days - weather forecasts
b) Forecasting the next few months - seasonal climate forecasts
c) Projecting the climate in the years to come - long-range climate forecasts (Blench, 1999).

Fig. 4.1: Local Traditional Knowledge (LTK) as embedded in knowledge-practice-belief Complex

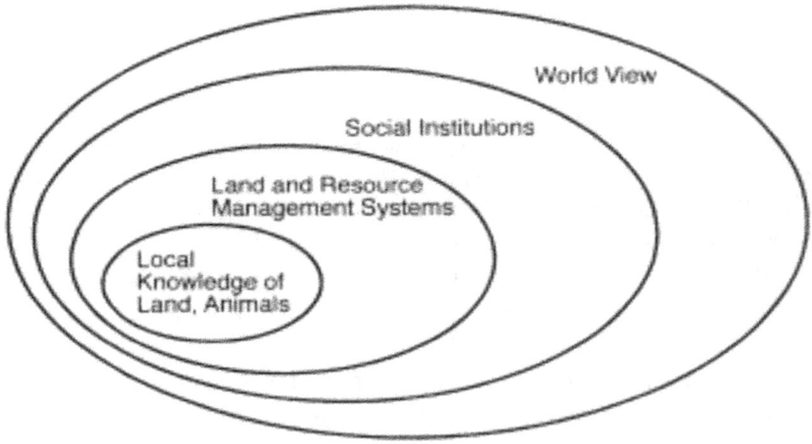

Source: (Berkes *et al.*, 2000)

LTK is outside the formal scientific domain, it is held by people in a particular geographic location (Tanyanyiwa and Chikwanha, 2011). It is often used interchangeably with indigenous or traditional knowledge, many terms, and definitions, for example, local knowledge (LK) indigenous knowledge (IK), traditional knowledge (TK), indigenous traditional knowledge (ITK), local ecological knowledge (LEK) and traditional ecological knowledge (TEK). LTK is knowledge based on experience, it is informal, tacit, dynamic and evolving, and is embedded in cultures that are socially differentiated. It incorporates practices (social institutions) as well as beliefs and value systems of a given society.

Indigenous Knowledge Systems (IKS) Weather Forecasting in Zimbabwe

Generally, in Zimbabwe's rural areas wild fruits such shuma (*Dyspros mespiliformis*), hakwa (*strychnosinnocu*) and nhunguru (*Flacourtia indica*) indicate drought or normal season by the amount of fruit they bear. If fruits are available in large quantities, this is an indication of drought whereas if there are few fruits a normal rainfall season is usually predicted. Pointers like the singing of the cicadas, crickets, rain cuckoo and appearance of migratory birds are witnessed during the early season. Old and new moon phases are associated with wet spells. IKS varies depending on region and community hence for the national level there is a need for comprehensive documentation of all IKS indicators. Young farmers need to be aware of the environment and use forecasts to plan on how to plant and when.

The Role of Local Traditional Knowledge: Key Challenges

There is lack of systematic documentation, coordination in the use of IK in weather and climate prediction (Tanyanyiwa and Chikwanha, 2011; Makwara, 2013). Generally, IKSs are localized and are in tandem to the farmers' context. However, this body of knowledge is threatened by urbanization, population growth, and climate change. The benchmarking of what is normal is difficult to comprehend in the use of IKS, e.g., how many swallows should reflect that it will be seasonal with normal rainfall. There is also the danger of the over-glorification of IKS indicators used in weather prediction. These are localised and based on oral communication which is difficult to verify thus limiting the applicability of IKS over large and heterogeneous areas (IDRC, 2010).

The conundrum is whether LTK was more relevant in the past than it is now, is it still relevant anymore or more relevant as the climate changes? In Zimbabwe, there is a general trend of looking down upon LTK mainly due to differences in how LTK is viewed as well as inputs from formal science (observations) that seem to be more appealing. Understanding the social context of an area is critical in understanding LTK. In other words, what and how to 'upscale' local experiences and use them outside areas of origin is a critical issue in understanding LTK as there are professional, institutional and political issues related to this type of knowledge. In Zimbabwe, current structures seem to favour formal science (education, governing institutions). This seems to be the government's main thrust as evidenced by its economic interests through power relations that include governance and land rights.

Integration of IKS and SCFs

Both IKS and seasonal climate forecasts (SCFs) have advantages and limitations. The major challenge is how to bring these two forms of knowledge together in a way that respects their values while building upon their respective strengths (IDRC, 2010). Forecasts may be more useful if ways are found to integrate local knowledge, which has enabled generations to live through severe floods and droughts into current management decision-making strategies. Table 3 shows a comparison of IKS and SCFs, and it further implores that if factors such as humidity, temperature, and wind speed are responsible for changes observed or perceived by local people, then meteorological data from individual stations can be summarized into a mean value correlating with a derived index.

Table 4.1: Comparison of SCFs and IKS

Factor	Seasonal Forecasts	Indigenous Knowledge
How approached	Compartmental	Holistic
How communicated	Written, easy to translate & communicate, somewhat standardized terms.	Oral, in local language & needs to be interpreted
How taught	Lectures, theories, & observations useful for scientific interpretation & modelling	Observation, experience; observations are of importance to daily life
How explained	Theory, 'value free'	Spiritual, social values
Repeatability	Repeatable measurements, can use same monitoring equipment in different places	Possibly poor repeatability, cannot expect similar perceptions & experiences for same observation
Precision	More precise, errors more objective & amendable; results from repeated measurements	Perceptive, individual, errors difficult to evaluate, every individual has different perceptions, possible bias
Spatial coverage	Time & space sparse, interrupted time-series, depends on monitoring network	Dense in space & time, long term observations; every individual is an observer
Tool	Designed instruments, limited, focused, recorded	Many undefined instruments, unfocussed observation, mostly unrecorded; different sensors, word of mouth, passing of information through generations
Scale	Large scale, general, conceptual	Smaller scale, specific, practical

Source: Adapted and modified from Dewalt, 1994 & IDRC, 2010

Current climate forecasts and the way they are presented and disseminated are ineffective, at worst they can increase risk due to inappropriate action by the decision maker (with potentially disastrous consequences), particularly for resource-poor, risk-averse subsistence farmers who in the absence of any scientific climate knowledge, management of agricultural systems have grown to cope with climate variability, often resulting in a wealth of indigenous knowledge relating to climate and weather (Kurukulasuriya and Rosenthal, 2003). Integrating IKS into weather forecasting and climate change policies can lead to the development of effective adaptation and mitigation strategies that are cost-effective, participatory and sustainable.

Barriers to Information Use

Information-related barriers to effective climate change adaptation relate to the provision of information and its dissemination. The use of climate information is therefore constrained by factors that relate to the production of the information and also to the communication process within the social, political, economic environment as is discussed in the successive paragraphs. Other critical constraints include the metaphors, words, strategies, frames, and narratives used in climate information communication, these need to be simplified so that they became palatable to a peasant farmer. Even if the information is produced, coordinated and shared appropriately, individuals may not necessarily use information in ways that support effective adaptation. This could be due to capacity constraints (a lack of skills or resources) or cognitive constraints (Productivity Commission, 2012).

Apart from the barriers relating to the generation of weather and climate information, there are also constraints to its use. Jones *et al.*, (2016) group the constraints to the use of climate information into five categories which are:

- the disconnect between producers and users of climate information; limitations of climate science; financial and technical constraints; political economy and institutional constraints; and psycho-social constraints.

Availability of (and access to) weather and climate information

Though societies worldwide have been coping with and adapting to climate change and variability and do as a result possess useful local knowledge, there is a need for marrying local or indigenous knowledge with climate projections and impact assessments at different levels including at a more local scale so as to inform adaptation. For many developing countries, local models are few if they exist at all and while global models exist, Kropp and Scholze (2009) point out that they often produce results that are inadequate for use in local assessments. According to Kropp and Scholze (2009), local climates are influenced significantly by smaller-scale features and processes, such as mountains, forests or lakes, the heat-island effect of large cities, which are not represented in detail in global climate models due to the low resolution yet the challenge for weather and climate information providers has been to downscale and provide information that is relevant to the local farmer. As a result, many developing countries lack adequate, reliable information which is necessary to inform adaptation options at very local levels.

For Zimbabwe, the National Climate Change Response Strategy sums it all by saying, 'most of the equipment at the Meteorological Service Department is old and now costly to maintain. The shortage of and lack of spare parts to repair faulty equipment gravely affects operations of the Department. The constant breakdown of equipment often results in research being carried out using a course resolution and creates gaps in datasets. The department has also suffered a major brain drain' (GoZ, nd). All these issues significantly impact on the quality of weather and climate information and consequently on the farmers' decision making. The uncertainty of climate change impacts makes the provision of and climate information a challenge and a serious one for developing countries, this is worsened by political, financial, technological, infrastructural, and human resources capacity issues associated with the collection and analysis of the data.

Participants at the UNFCCC workshops and meetings reported that observations and data availability still need to be improved in all regions. At the Africa workshop, participants agreed that systematic observation networks in Africa are inadequate because there is a low coverage of weather stations and lack of maintenance of the existing ones. This has an impact on the quality of forecasts and consequently the credibility of national meteorological stations. Participants reiterated that missing and scattered observational climate data in Africa is a constraint to understanding current and future climate variability. If data exists, there are difficulties in obtaining it (UNFCCC, 2007).

A study which was done by Mawere *et al.*, (2013) in Chivi and Zvishavane Districts in Zimbabwe indicated that though farmers were aware of the changes in local weather and climate patterns, most farmers (52%) indicated that they had no formal means of accessing information about current weather conditions and had to rely on the traditional knowledge systems (IKSs), which include prophecy, general observations and body feelings. The farmers indicated that they rely on observations of flora and fauna which is a common thing in the rural areas as very few have access to newspapers and radio weather forecasts (Mawere *et al.*, 2013).

Effective adaptation planning requires improved observations; improved regional, national and global data, as well as denser networks; the recovery of historical data; the building of support among the user communities that have a demand for climate information; and promoting greater collaboration between the providers and users of climate information (UNFCCC, 2007). This is particularly important considering that seasonal forecasts and predictions are perishable and as such data collection, analysis, and communication to the stakeholders has to be ongoing and timeous.

Typical Weather Forecasts in Zimbabwe

Weather forecasts that are given in Zimbabwe assume that everyone understands English well (Fig. 14). The assumptions of this type of communication are that the public will be able to understand the meaning.

Fig. 4.2: A Typical Weather Forecast in Zimbabwe

ZIMBABWE

MINISTRY OF ENVIRONMENT, WATER AND CLIMATE

Meteorological Services Department

Telephone: +263 (04) 778 206 **Fax:** +263 (04) 778 161

Website: www.weather.co.zw

P.O Box BE 150, Belvedere, Harare, Zimbabwe

All correspondences should be addressed to the Director

MSD

METEOROLOGICAL SERVICES DEPARTMENT

Minimizing Risks through Science

20 JUNE 2017

ZIMBABWE EVENING WEATHER FORECAST ISSUED AT 1600 HRS ON TUESDAY 20 JUNE 2017 VALID UNTIL THURSDAY 22 JUNE 2017

The cloud band currently affecting the much of Zimbabwe should maintain moist conditions over the country. These weather conditions are therefore expected to cause cloudy spells now and again in most areas.

Tomorrow Wednesday 21 June then, in all 3 **Mashonaland** provinces, **Harare Metropolitan** province, north of **Midlands** province as well as **Manicaland** province, we expect it to be cloudy and rather cold with sunny breaks from time to time. In the early morning there are high chances of frost occurring mainly in Manicaland province (Nyanga, Rusape as well as Mutare) and Mashonaland East province.
To the south, that is **Masvingo**, south of **Midlands** province (stretching from Gweru through Mberengwa to Zvishavane) and **Matabeleland South** provinces, it is also expected to be cold in the morning. As the day progresses, cloudy, mild and windy weather will prevail.
Although slight ground frost may occur in the early hours of the morning, it should be confined to areas like Kezi and Matopos.
In Matabeleland North province, it should be generally clear and cold in the morning becoming mild later with some occasional cloudy periods.

Thursday 22 June:
The weather should continue cloudy and cold and rather windy with sunny breaks in all areas.

Recommendations

These cold weather conditions are a recipe for the spread of colds and flu, therefore it is incumbent upon us all to put on warm clothing at all times.
Braziers should be used with care to avoid carbon monoxide poisoning

TEMPERATURES FOR THE MAIN CITY CENTRES (°C):

	Harare	Bulawayo	Gweru	Mutare	Kwekwe	Kadoma	Masvingo	Chinhoyi	Lupane	Bindura	Marondera	Gwanda
Max today	20	19	19	24	21	21	21	23	24	23	19	23
Min tomorrow	05	12	05	05	10	10	05	07	12	07	02	05
Max tomorrow	18	19	18	22	19	20	19	22	23	21	17	21

ISO9001:2008 Certified

@MSDZIM Meteorological Services Department of Zimbabwe +263 778 125 911

Source: Meteorological Services Zimbabwe, 2017

However, this type of communication fails to ascertain the targeted audience and selecting a credible messenger for that audience. In the same vein, the type of claim, argument asserted and why it is appropriate for the audience is not clear. The message that is given is not connected to socio-cultural values and beliefs that is, the local indigenous knowledge systems for a particular area. Basically, people react to traditions, experiences and shared values as compared to abstract concepts and scientific data. The message is not meaningful to the less educated, that is, it does not appeal to values that are meaningful to the local audience. A good example would be speaking in spiritual language and parables when targeting a conservative Christian audience. The strongest argument is not spelt out because the message is not empowering as specific actions to be taken to make a difference are not outlined. The notion of globalization is not fostered that is, there is no link to global patterns, collective action and actions as well as other organizations, key players, leaders, etc. because actions and events are an effective mode of communication.

Effectiveness of Communication

Apart from the availability and accessibility of information, how the information is communicated is equally important. Roger, Pulwrty, Segun, and Zorba (2010) argue that for framing effective communication the following questions have to be answered:

a) Is the information relevant to decisions in the particular agricultural system?
b) Are the sources/providers of information credible to the intended user?
c) Are farmers receptive to the information and to research?
d) Is the research accessible to the policymaker or decision-maker?
e) Is the information compatible with existing decision models and farming practice?
f) Do decision-makers have the capacity to use information?

The questions raised by Roger, Pulwrty, Segun and Zorba (ibid) indicate that factors such as credibility of the sources, methods of communication (accessibility), the relevance of the information to the user and the ability of farmers to use the information all affect the use of weather and climate information.

Channels of Communicating Weather and Climate Information

The communication channel used to disseminate climate and weather information influences the uses of the information. According to Jones *et al.*,

(2016) existing dissemination channels for climate information often do not reach decision makers and communities equally. Sometimes information may reach communities when it is no longer necessary. For effective communication of weather and climate information, according to Roger *et al.*, (2010), there is need to identify various channels of communicating weather and climate information which include:

a) Workshops and meetings (shared scenario construction, shared model building);
b) Presentations and briefings (including locally organized events, for example, hearings);
c) One-on-one technical assistance;
d) Coordination with other on-going projects;
e) Work with the local media;
f) Website development and maintenance;
g) Courses on climate impacts and adaptation
h) Media (mass media and information, telenovelas (soap operas) and the like)

For the communication of climate information to be effective a trusted communicator should be used; the trusted communicator should be someone whom the farmers already know and trust (Roger *et al.*, 2010). Existing institutions in Zimbabwe include the agricultural extension department or some NGOs which have a history of working with the farmers. More to that, among other issues, the issue of the language used for communication may need to be considered especially in developing countries where languages inherited from the colonial masters may not be sufficient for different information users. If local languages are used together with relevant channels of communication (for example local dramas, songs, etc.), the majority of recipients are more likely to understand.

Trust and Credibility of the sources

Credibility is an issue of trust and belief and is the degree to which actors believe the information as true, and hence are willing to rely on it. It is a function not only of the message but also of the messenger. Information users may not trust the information if past forecasts appear to have been wrong. According to Patt (2001), the credibility constraint arises both from the failure of past years' forecasts to be accurate, and from the failure of the forecast communicator to have built a reputation of trust. In Zimbabwe, farmers receive the official forecast through the media (radio and newspapers), and other sources such as agricultural extension officers. Patt (2001) argued that though the media may be the most effective way of informing a large number of people, it is unlikely to increase the trustworthiness of the forecasts. Patt's argument was that the media in Zimbabwe are 'primarily state-run and perceived as serving the interests of the governing political party' and therefore lack credibility when it comes to climate change information communication. However, at the moment there are other privately run newspapers, but the source of information remains the same. While the public may not trust the state-run newspapers the question that needs to be asked is can they trust the Meteorological Services Department? The issues relating to the credibility of the Meteorological Services Department in Zimbabwe may be a result of a myriad of factors including poor infrastructure, the use of deterministic (instead of probabilistic) as well as failure to downscale and provide information that is relevant to the local level users. Zimbabwe's Meteorological Services Department is, therefore, under pressure to produce reliable information so as to adequately and reliably inform farmers.

Cognitive Constraints

Cognitive constraints also influence the communication method used and the type of information used. Due to the thinking or idea that understanding a probabilistic forecast by the public may be difficult, in Zimbabwe and many other developing countries, forecasts are communicated as being deterministic, rather than probabilistic. In other words, those responsible for communicating the weather and climate information do not trust the mental capacity of farmers to understand probabilistic forecasts, and in the end, the credibility of Zimbabwe National Meteorological Office has been questioned. Thus there is a communication challenge in translating science into practical options and guidance (Jones *et al.*, 2016).

As noted above, many providers of weather and climate information choose to communicate deterministic forecasts because they are worried that farmers cannot understand the probabilistic versions. If users do not understand a forecast, it is not likely that they will use it, and if they do use, they may use it wrongly. According to Patt (2001) 'simplifying the forecast is an attempt to circumvent the cognitive constraint' though it may yield unintended results. There is, therefore, need for climate scientists to match the information needs of farmers and other decision makers so that the information remains relevant and usable.

Risk Perceptions

According to IPCC (2007) knowledge of climate change causes, impacts and possible solutions does not necessarily lead to adaptation and providing individuals with scientifically sound information may not provide the anticipated results. Risk perception of individuals will have an effect on an individual's response as Dang, Li and Bruwer (2012) point out that 'human behaviour is more complex, and human actions are likely to be motivated by psychological factors such as personality, belief, perception or habit' risk perception will influence the response to weather and climate information of decision maker (institution or individual). According to Jones *et al.*, (2016) risk perception is a key driver of institutional and political change, and low levels of perceived risk can, therefore, contribute to inaction or the prioritization of other risks ahead of climate adaptation.

Lack of Data

The Meteorological Services Department is mandated by national laws and recognised in the Convention of the World Meteorological Organization. Its aim is to collect and serve meteorological and hydrological forecasting and warning systems at country level. The MSD operates well below capacity due to underfunding and low visibility. Regrettably, about 80% of the data from the meteorological service is unable to provide proper climate information and early warnings due to decades of neglect. In the same vein, the World Bank estimates that about USD$1 billion is needed to modernise Africa's meteorological services. Zimbabwe does not have the capacity and expertise to provide complete solutions to its meteorological services which challenges Partnerships in the agriculture, insurance and telecommunication sectors for the collection and the delivery of data and for critical services including risk analysis, commodity prices, insurance, and secure payment schemes are important. Innovative solutions include that by ECONET, a local mobile phone operator in Zimbabwe. It started a large scale weather-indexed insurance for farmers in Zimbabwe, known as Ecofarmer which has benefited more than 900,000 farmers. Zimbabwe needs Strong political support to increase smart systems through partnerships with national authorities, technical agencies, NGOs and the private sector including investing in modernising weather and climate data.

Conclusion and the Way Forward

In Zimbabwe, farmers cannot access reliable and usable weather data. Information is often unavailable. If it does exist, the quality is poor, or it is inaccessible to those who need it most. Farmers do not get reliable information on drought forecasts, rainfall distribution, and pest outbreaks. The Zimbabwe government and development agencies do not recognize and prioritise the value of climate and weather data. This has stifled investment in infrastructure and proper functioning of state institutions charged with collecting and serving climate and weather data, the Meteorological Services Department, Zimbabwe. Zimbabwe can tap into the USD$10.2 billion Green Climate Fund adopted in 2011 by the UNFCCC to finance adaptation and mitigation projects and programmes in developing countries. Financing infrastructure on basic meteorological observations to generate climate and weather data fit for agricultural purposes should be a priority for the fund. Though many players are involved in the generation and dissemination of weather and climate information, the Zimbabwe Meteorological Service Department remains the main source of seasonal weather forecasts in Zimbabwe. Economic and technological challenges have impacted on the generation and dissemination of weather and climate information. In Zimbabwe, there has been a mismatch between farmers' needs and the disseminated information which is usually communicated in English and as such farmers have not been able to use the information in the decision-making process effectively.

Based on the conclusions made by this study, the authors make the following recommendations:

- Improvement in stakeholder involvement in the collection and dissemination climate information
- Capacitating Zimbabwe Meteorological office to give clear-cut information that is relevant to all stakeholders especially smallholder farmers' who live at the margins
- There is a need for inter-institutional collaboration (and multi-stakeholder involvement) among national meteorological and hydrological services and agencies such as extension services, development projects, and community-based organizations and non-governmental organizations (NGOs) in as far as of weather and climate information for effective adaptation is concerned.
- Capacity building, teaching users what the information means, and how to interpret the complexity of weather and climate information is required.

References

Adger, WN, S. Agrawala, M.M.Q. Mirza, C. Conde, K. O'Brien, J. Pulhin, R. Pulwarty, B. Smit & K. Takahashi, 2007, Assessment of adaptation practices, options, constraints, and capacity. Climate Change 2007: Impacts, Adaptation, and Vulnerability. Contribution of Working Group II to the Fourth Assessment Report of the Intergovernmental Panel on Climate Change, M.L. Parry, O.F. Canziani, J.P. Palutikof, P.J. van der Linden & C.E. Hanson, (Eds.), 2007, Cambridge University Press, Cambridge, UK, 717-743

Brazier, A. 2015, Climate Change in Zimbabwe: Facts for Planners and Decision Makers. Konrad-Adenauer-Stiftung, Harare.

Berkes, F., J. Colding & C. Folke 2000, Rediscovery of Traditional Ecological Knowledge as Adaptive Management. Ecological Applications (10) 1251-1262.

Blench, R. 1999, Seasonal climate forecasting: Who can use it and how should it be disseminated? Natural Resource Perspectives (47) 1–5.

Dang, H., Li, E. and Bruwer J. 2012, Understanding Climate Change Adaptive Behaviour of Farmers: An Integrated Conceptual Framework: The International Journal of Climate Change: Impacts and Responses (3) 2.

Dewalt, B.R. 1994, Using Indigenous Knowledge to Improve Agriculture and Natural Resource Management. Human Organization 53 (2), 123-131

Government of Zimbabwe (n.d) Zimbabwe's National Climate Change Response Strategy

Ministry of Environment and Natural Resources Management, Harare.

Government of Zimbabwe, 2013a, Zimbabwe National Climate Change Response Strategy (First Draft). Ministry of Environment and Natural Resources Management, Harare.

Githungo, W. N., Kinuthia, R. G., Kizito, K. & Rao, K. P. C., 2010, Effects of Seasonal Forecasts on Farm Decision Making for Smallholder Farming systems in Semi-Arid Parts of Kenya. Proceedings of the 2nd International Conference on Climate, Sustainability and Development in semi-arid regions, ICID, held in Brazil in August 2010.

IDRC, 2010, Integrating meteorological and indigenous knowledge-based seasonal climate forecasts for the agricultural sector: Lessons from participatory action research in Sub-Saharan Africa. Climate Change and Adaptation in Africa (CCAC).

IPCC, 2007, Climate Change, 2007, Impacts, Adaptation and Vulnerability. Contribution of Working Group II to the Fourth Assessment Report of the IPCC, Cambridge University Press.

Jones, L., Champalle, C., Chesterman, S. Laura Cramer & Crane T.D. 2016, Constraining and enabling factors to using long-term climate information in decision-making. Climate Policy (17), 551-572.

Konneh, K.A., 2006, An Assessment of NOAA Funded Agriculture and Climate Interactions Research 1995–2005. NOAA Climate Program Office Report (draft). NOAA, Silver Spring.

Kropp, J, and Scholze M, 2009, Climate Change Information for Effective Adaptation: A Practioner's' Manual, (GTZ) Eschborn/Germany.

Lu, X, 2009, Applying Climate Information for Adaptation Decision-Making: A Guidance and Resource Document. New York: National Communications Support Programme, UNEP, GEF.

Kurukulasuriya, P. & Rosenthal, S. 2003, Climate Change and Agriculture: A Review of Impacts and Adaptations. Paper No. 91 in Climate Change Series, Agriculture and Rural Development Department and Environment Department, World Bank, Washington, DC.

Madhavan, M & Rengalakshmi, R, ND, Seasonal Climate Information for Ensuring Agricultural Sustainability and Food Security of Small Holder Rainfed Farmers: Experience from India, Available from: http://www.gdn.int/fullpaper/Session-18-Manjula.M%20and%20 Rengalakshmi.R-Seasonal%20Climate%20Information.pdf Accessed [13/02/17]

Makwara, E. C. (2013). Indigenous Knowledge Systems and Modern Weather Forecasting: Exploring the Linkages. Journal of Agriculture and Sustainability 2 (1):8-141.

Manatsa, D., Unganai, L., Gadzirai, C. and Behera, S.K., 2012, An Innovative Tailored Seasonal Rainfall Forecasting Production in Zimbabwe. Natural Hazards (64), 1187–1207.

Mason, S.J., Joubert, A. M., Cosijn, C. and Crimp, S. J. 1996, Review of seasonal forecasting techniques and their capabilities. Water SA 22 (3).

Mawere, M., Madziwa, B.F., & Mabeza, C.M., 2013, Climate Change and Adaptation in Third World Africa: A Quest for Increased Food Security in Semi-Arid Zimbabwe. The International Journal of Humanities & Social Studies, 1(2).

Meza, J.F., James, W., Hansen, J.W., & Osgood, D. 2008, Economic Value of Seasonal Climate Forecasts for Agriculture: Review of Ex-Ante Assessments and Recommendations for Future Research. Journal of Applied Meteorology and Climatology, (47), 1269-1286

Meteorological Services Department, 2017, Harare, Zimbabwe

Mupakati, T. & Tanyanyiwa, V.I., 2017, 'Cassava production as a climate change adaptation strategy in Chilonga Ward, Chiredzi District, Zimbabwe', Jàmbá: Journal of Disaster Risk Studies 9(1), a348. https://doi.org/10.4102/jamba.v9i1.348). 1996-1421

Patt, A., 2001, Helping Farmers in Zimbabwe Use Seasonal Climate Forecasts: Six Constraints to Effectiveness: Paper presented at the Open Meeting of the Human Dimensions of Global Environmental Change Research Community, Rio de Janeiro, Brazil, 6-8 October, 2001, http://sedac.ciesin.org/openmeeting/downloads/1005784679_presentation_apatt_riopaper1.pdf

Productivity Commission 2012, Barriers to Effective Climate Change Adaptation, Report No. 59, Final Inquiry Report, Canberra.

Productivity Commission 2012, Barriers to Effective Climate Change Adaptation, Report No. 59, Final Inquiry Report, Canberra.

Roger S., PulwartyR. S., Segun O.O., & Zorba, P., 2010, Communicating Agroclimatological Information, Including Forecasts, for Agricultural Decisions. In: WMO (2010) Guide to Agricultural Meteorological Practices, World Meteorological Organization, Geneva.

Suarez P., 2009, Linking Climate Knowledge and Decisions: Humanitarian Challenges. The Pardee Papers/No. 7 Boston University http://www.bu.edu/pardee/files/2010/01/Pardee_Paper__7_Linking_Climate_Knowledge.pdf.

Tanyanyiwa, V.I. and Chikwanha, M, The role of indigenous knowledge systems in the management of forest resources in Mugabe area, Masvingo, Zimbabwe. Journal of Sustainable Development in Africa. 3(3). 132-149

UNICEF & IES, 2014, Children and Climate Change in Zimbabwe https://www.unicef.org/zimbabwe/Children_and_Climate_Change_in_Zimbabwe_Report_2014.pdf

UNDP, 2011 Africa Adaptation Programme: Capacity Building Experiences Improving Access, Understanding, and Application of Climate Data and Information. UNDP Environment & Energy Group, Discussion Paper Series, 2, New York.

UNFCCC, 2007 Climate Change: Impacts, Vulnerabilities and Adaptation in Developing Countries, United Nations Framework Convention on Climate Change, Bonn, Germany.

Vermeulen, S.J., Aggarwal, P.K., Ainslie, A., Angelone, C., Campbell, B.M., Challinor, A.J., Hansen, J., Ingram, J.S.I., Jarvis, A. & Kristjanson, P., 2010, Agriculture, Food Security and Climate Change: Outlook for Knowledge, Tools and Action. CCAFS Report 3. Copenhagen, Denmark: CGIAR Research Program on Climate Change, Agriculture and Food Security

World Meteorological Organization, 2010, Guide to Agricultural Meteorological Practices, World Meteorological Organization, Geneva.

Yeboah, S., 2017, How a lack of access to reliable weather data is hurting African farmers available at https://theconversation.com/how-a-lack- of-access-to-reliable-weather-data-is-hurting-african-farmers-. Accessed on 08/07/2017

Ziervogel, G. & T. Downing, 2004, Stakeholder networks: Improving seasonal forecasts. Climatic Change, (65), 73–101.

Ziervogel, G., 2004 Targeting Seasonal Climate Forecasts for Integration into Household Level Decisions: The Case of Smallholder Farmers in Lesotho. Georg. J., (170), 6–21.

Ziervogel, G., & Calder, 2003, Climate Variability and Rural Livelihoods: Assessing the Impact of Seasonal Climate Forecasts in Lesotho. Area 35 (4), 403– 417.

Chapter Five

PERCEPTION OF CLIMATE CHANGE IN NEPAL

Anup K C, *Tribhuvan University, Kathmandu, Nepal*

Abstract

Climate change is a burning issue which affects different aspects of the environment. Nepal is one of the most vulnerable countries to climatic impacts due to its diverse topography. Temperature rise and shifts in rainfall pattern have caused adverse impacts in different sectors. There is a severe impact of climate change on glacial ecosystem and agricultural productivity. The Government of Nepal has focused more on developing policies and less towards implementing climate change adaptation measures. This paper focuses on a review of climate change research throughout Nepal. Many researchers focus on a questionnaire survey, field observations, key informant interviews, focused group discussions, climatic data collection, and modelling. People perceive an increase in temperature, change in rainfall trend, decrease in snowfall and melting of glaciers due to climate change. It has an adverse impact on agriculture, biodiversity, livelihoods, hydropower, water resources and tourism. It is recommended to install accurate climatic data collection system and forecasting, effective early warning system, vulnerability assessment and awareness generation to minimize risks of climate change. Also, climate change adaptation measures should be focused on enhancing economic, social and ecological sectors in Nepal.

Key words: Climate Change • Multiple Sectors • Nepal • Perception •Scenario

Introduction

Climate change is a global, regional and local issue which is affecting humanity and ecosystems (K C et al., 2016; Acharya and Bhatta, 2013). It is not a new issue as it has been occurring historically (K C and Ghimire, 2015). However, the impact has now risen due to the anthropogenic emission of greenhouse gases (GHGs) (K C et al., 2013; K C et al., 2014). There has been more than a 0.6°C rise in global temperatures since 1700 (Dahal et al., 2009), while the IPCC (2007) analyzed temperature rise of 0.74°C in the hundred years' interval (1906-2005) (K C and Thapa Parajuli, 2014). Severe impacts of climate change have been observed in the Hindu Kush Himalayan region including Nepal (K C and Ghimire, 2015; Shrestha et al., 1999). Increase in global temperature by 2°C would disrupt the climate system and biodiversity. There will be an intense warming of the earth's surface leading to global water scarcity. It will also result in warm summer days, a change to rainfall pattern, more rainfall extremes and droughts, destruction of high altitude ecosystems of Asia, glacial melt, sea level rise, forest degradation and the shifting of the tree line and biodiversity loss (Dahal *et al.*, 2009). According to the IPCC, 25% of the global mountain glaciers may disappear by 2050.

Climatic Scenario in Nepal

Nepal is a country with diverse climate and biodiversity due to its high altitudinal range of 60m asl in the south to 8,848m asl in the north (ADB, 2015; K C, 2015). The annual mean temperature of Nepal is about 15^{0}C which decreases from the south to the north (Alam and Regmi, 2004). Average annual rainfall is about 1,800mm, varying from 5,000mm in Lumle to 250mm in the rain shadow region near Tibet (NCDC, 2010; Alam and Regmi, 2004). The contribution of summer monsoon rainfall is more than 80% of the total annual rainfall (Baidya *et al.*, 2007; NCDC, 2010). The climate of the country is mainly divided into four different seasons. The summer monsoon, which consists of cloud and rain, starts in June and ends in September. There is variation in magnitude and intensity of rainfall in different locations due to high altitudinal variation and rugged topography. The magnitude of summer monsoon rainfall decreases from the southeast to the northwest. Post-monsoon rainfall starts in October and ends in November with small spells of thundershowers. Temperatures decrease slowly during this season. The winter season, with its clear dry skies and scanty rain, starts in December and ends in February. Winter monsoons decrease in magnitude from the north-west to the southeast. It is the coolest season with minimum temperatures in January. The premonsoon rainfall with lightning and thunderstorms in the hilly areas starts in April and ends in May. It is the hottest season with the highest temperatures in May which gradually decreases after the start of rainfall (NCDC, 2010). The climate of Nepal varies from tropical to arctic from the south to the north within a short distance of 200km only (Alam and Regmi, 2004).

Climate Change in National Scenario

Climate change affects the environment and society. The effects of climate change have been felt in all parts of Nepal in the form of an increase in temperature, changing rainfall patterns and frequent disasters (Regmi *et al.*, 2010). Average temperatures are increasing at the rate of 0.04-0.06°C per year (Shivakoti *et al.*, 2015). There is a high increase in temperature during dry winters and less during monsoon (Agrawala *et al.*, 2003). There are more frequent hotter days and less frequent cooler nights (Shivakoti *et al.*, 2015). The summer season is getting warmer while the winter season is getting cooler (Regmi *et al.*, 2010). Different climatic models have forecasted an increase in temperature for the years 2030, 2050 and 2100 in Nepal. There is a forecast of an overall increase in temperature of about 1.2°C till 2050 and 3°C till 2100 (Agrawala *et al.*, 2003).

Projections of rainfall are less certain than that of temperature (Alam and Regmi, 2004). There is an increasing trend of monsoon precipitation throughout Nepal with the exception of a few mid-western, southern, central and eastern locations. There is also an overall increase in monsoon precipitation by 15 to 20%. Increasing trend of rainfall is higher in eastern and central Nepal than compared to western Nepal (Shivakoti *et al.*, 2015). Intensive monsoonal rainfall events accompanied by flooding and irregular discharge are on the rise (Agrawala *et al.*, 2003; Shivakoti *et al.*, 2015). Far-west Nepal and lower altitude parts in the east, west, and centre show an increasing trend of post-monsoon precipitation while higher altitudes in the west, centre, east and far-west show a decreasing trend. Except for the higher altitude of mid-west and the east, winter precipitation shows an increasing trend. There is less variation of winter precipitation in western Nepal, but there is a 5 to 10% increase in the eastern part (Shivakoti *et al.*, 2015). There is also the less frequent occurrence of droughts. Different climatic models predict an increase in precipitation in the future (Agrawala *et al.*, 2003).

Climate change adversely affects biological, physical, chemical, socio-cultural and economic aspects of environment due to its diverse topography (Parajuli, 2010). There is an adverse impact on infrastructure, agriculture, drinking water, irrigation, hydropower, and biodiversity (Alam and Regmi, 2004). It also affects forestry, energy, and health (Regmi et al., 2010). Nepal is a less-developed country with high vulnerability and less adaptive capacity due to its high mountains and the Himalayas. Climate change has direct as well as indirect adverse impacts on the livelihood of people. Poor and agriculture-dependent households are suffering more from severe droughts and floods (Dahal et al., 2009). Many people are migrating to cities and flying abroad to cope with the adverse impact of climate change (Regmi et al., 2010). Poverty and food insecurity are increasing due to low agricultural productivity, as a result of climatic hazards (Shivakoti et al., 2015). It is further increasing food shortage and food prices (Regmi et al., 2010). Women, children, the poor, marginalized and smallholder farmers are unable to cope with the changing climate and associated hazards. Farmers are focusing on vegetables, low water requiring crops, changing farming practices, rainwater harvesting and erosion control (Shivakoti et al., 2015).

People of Terai region are experiencing unexpected monsoon flash floods, cold waves, winter droughts and long-term dry spells in annual precipitation (Regmi et al., 2010; Agrawala et al., 2003; Shivakoti et al., 2015). Change in hydrological pattern and increased glacial retreat in spring and summer affects irrigation systems, water mills, hydropower plants and drinking water supply systems. It results in a decrease in evaporation and agricultural production; and an increase in malnutrition, water-borne disease, and sediment load. It also causes an adverse impact on health, sanitation, human settlements, infrastructure, agricultural land, mountain ecological zones and scarce fuel-wood (Agrawala et al., 2003; Shivakoti et al., 2015). Increase in forest fires, dry spells, pests, and diseases are also observed as a result of climate change (Regmi et al., 2010). Climate change causes a shift in dominant species of the lower region to higher region (K C and Ghimire, 2015).

The contribution of Nepal to global warming is very low, but adverse impact of global warming in glacier ecosystem is very high (GoN, 2015). Rapid melting of glacier leads to the development of glacial lakes in recent decades (Shivakoti et al., 2015). There is less snowfall resulting in the retreat of glaciers in the Himalayan region of Nepal (Regmi et al., 2010). Climate change is increasing environmental, social, political and economic issue. The vulnerability of Nepal towards climate change and natural disasters is increasing due to the adverse change in amount, intensity, frequency and form of precipitation (GoN, 2015). Also, the water resources in the hilly regions are shrinking (Regmi et al., 2010).

Climate Change Perception in Policy Level of Nepal

Different governmental and non-governmental organizations are involved in the climate change sector in Nepal. They are using different mediums for climate change communication in a different part of the country (K C, 2017). The Government of Nepal (GoN) is very strong in climate change related policy formulation and legislation, but it is very weak in its implementation. The Ministry of Population and Environment (MOPE) is the focal point for the United Nations Framework Convention on Climate Change (UNFCCC) which coordinates climate-related activities in the country (Alam and Regmi, 2004; MoPE, 2016a). Nepal has been a part of the UNFCCC since July 1994 and Kyoto Protocol since December 2005 (CCNN, 2011; CEN, 2012). Ministers of Nepal attended the Conference of the Parties (COP) only after 2008 (CEN, 2012). To address the negative impacts of climate change, COP 7 in 2001 focused on preparing and implementing National Adaptation Programmes of

Action (NAPA) (MoSTE, 2015c). Preparation and endorsement of NAPA on 28 September 2010 by GoN was an important step towards climate change adaptation (FAO, 2010; GoN, 2011). NAPA set 43 adaptation options in 9 priority areas (ADB, 2015). There was an allocation of 80% budget at the Village Development Committee (VDC)/Municipality level through a line ministry and District Development Committee (DDC) (CCNN, 2011). The National Framework on Local Adaptation Plan for Action (LAPA) was a bottom-up, inclusive, responsive and flexible framework implemented in the 14 most climate vulnerable districts (Maharjan, 2014). It focused on aspects of adaptation and resilience in local and national plans (MoE, 2011). The Pilot Programme for Climate Resilience (PPCR) aimed to increase resilience and adapt to climatic risk in the infrastructural sector with the support of Asian Development Bank (ADB) (MoSTE, 2015a).

Nepal Climate Change Support Programme (NCCSP) supported the poorest and most vulnerable communities and helped GoN to implement Climate Change Policy and LAPA from 2013 to 2015. It implemented 70 LAPAs in the 14 most climate vulnerable districts of far and mid-west region with the support of Department for International Development (DFID), European Union (EU) and the Cyprus Government (Maharjan, 2014; MOPE, 2016c). It aimed to integrate climate change into national development agenda and contribute to poverty reduction, livelihood diversification and community resilience (MoPE, 2016c). Hariyo Ban Program was a five-year USAID funded program implemented in 2011 by World Wide Fund for Nature (WWF) Nepal, Cooperative for Assistance and Relief Everywhere (CARE), National Trust for Nature Conservation (NTNC) and the Federation of Community Forestry Users Nepal (FECOFUN). It focused on biodiversity conservation, sustainable landscape and climate change (Maharjan, 2014). GoN prepared Strategic Program for Climate Resilience (SPCR) with the support of ADB, World Bank and other stakeholders (ADB, 2015).

COP 16 decided to plan and support National Adaptation Planning (NAP) process in the least developed countries (LDCs) to implement adaptation programmes (MoSTE, 2015c). Guidelines on NAP were prepared by UNFCCC to prepare adaptation program and modify policies to make LDCs more resilient. As being a party to the UNFCCC, Nepal committed to preparing short-term, mid-term and long-term climate change adaptation plan including National Adaptation Plan (NAP) in September 2015 with the support of DFID and Practical Action (MoSTE, 2015c). Implementation of community forest management programmes, promotion of hydropower and prioritizing the use of other clean energy sources by GoN is helping towards climate change mitigation in Nepal (Sharma, 2009). Seven different programmes on climate change were proposed to reduce the climatic risk of the people and communities of Nepal by improving the performance of public and private sector institutions. Sixty-six project level indicators inside five core indicators specific to Nepalese context were taken for addressing issues of climate change resilience building. It covered seven sectors and eight districts and provided a rich resource of information to inform policy and planning reforms (MoSTE, 2015d). Climate change adaptation planning in agriculture focused on increasing resilience towards drought, flood, disease, pest and climate change and promote conservation agriculture (MoSTE, 2015d).

Methodology

This chapter is based on secondary information on climate change perception. It is fully based on literature review of climate change research and climate change documents available in Nepal and internet sources. The literature on climate change perception was available from Humla, Pyuthan, Gorkha, Mustang, Kaski, Manang, Myagdi, Lamjung, Kathmandu, Sindupalchowk, Chitwan, Taplejung, Panchthar, Ilam, Palpa, Jumla, Dolakha, Kailali, Bhaktapur, Baglung, and Banke districts. The studied districts lie in eastern, central and western Nepal (Fig. 5.1). Different researchers on climate change perception in different regions use different methodologies. There is a need for a review of methods which would help the researchers to find the subject of research with appropriate methodology. It would be easier and faster to select appropriate literature necessary for them.

Fig. 5.1: Distribution of studied Districts (Nepal)

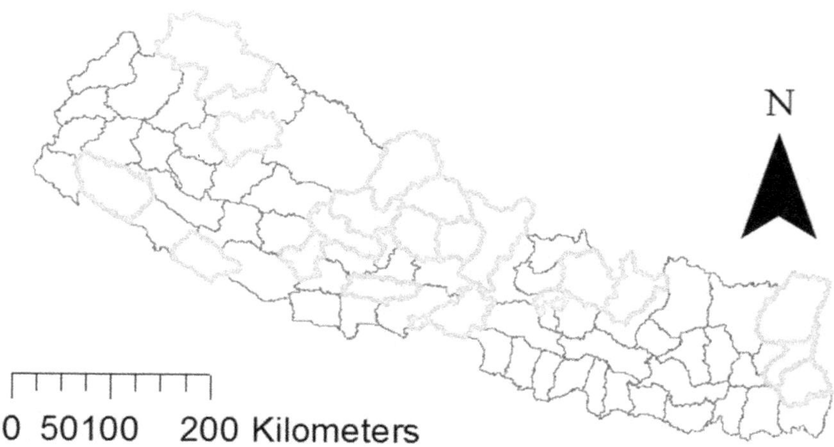

DOE (2009) conducted a study in Northern Ireland to assess the perception of people on climate change and livelihood. Different aspects of climate change and its impact were identified through questionnaire surveys. With the help of questionnaire survey, Ishaya and Abaje (2008) assessed the perception of climate change by indigenous people in the Kaduna State of Nigeria. With the help of four parameters in a scoring model, Ehmer and Heymann (2008) studied positive and negative effects of climate change in different countries. They examined direct effects of climate change as a first parameter, substitution effects of climate change as a second parameter, geographical substitution effects and regulation effects as a third parameter and adaptation to climate change as a fourth parameter.

To assess the impact of climate change on the agricultural gross domestic product (AGDP), modelling was done with the help of data of annual AGDP, average rainfall, temperature, seeds and fertilizer distribution in Nepal. The model uses the annual data of agricultural gross domestic product (AGDP), average rainfall, temperature, seeds and fertilizer distribution in Nepal. Data on AGDP was taken from Central Bureau of Statistics (CBS), data on seeds and fertilizer was collected from the Department of Agriculture and data of rainfall and temperature was available at the Department of Hydrology and Meteorology (DHM) (Acharya and Bhatta, 2013). To assess different indigenous practices and its role in climate change adaptation, primary data and information were collected from key informant interviews (KIIs) and household surveys while secondary information was collected through review of the literature, national and district level stakeholder workshops and focus group discussions (FGDs) (MoSTE, 2015b).

A qualitative approach was used by Onta and Resurreccion (2011) to assess the adaptation practices of people in Humla District of Nepal from June to December 2009. Observation, FGDs, and KIIs were used to collect the data. To get information about climate change adaptation and socio-cultural structures, interviews were conducted with 18 respondents at the district level. With an objective to assess climate change adaptation measures at the local level, Regmi and Bhandari (2013) conducted a study in different regions of Nepal. The study was based on literature review, questionnaire survey, and case studies. Also, a semi-structured questionnaire survey was done with 17 policy makers and 26 practitioners working on climate change issues. The case study was conducted in two pilot VDCs of LAPA and Community Adaptation Planning (CAP) implementation in Pyuthan district of Nepal.

Lamichhane and Datt (2009) conducted a climate change perceptual study in Lundi Khola sub-watershed of Gorkha District, Nepal from April-May 2008. Reconnaissance surveys, KIIs, and analysis of climatic data were carried out for conducting research. Perception of people towards climate change was taken in Mustang District of north-west, Nepal while preparing district energy and climate change report. The study was conducted with the help of public consultations, KIIs, meetings, and FGDs. Checklists were prepared to take information from local people, governmental staffs and local non-governmental organization representatives (CEPAD, 2011). To assess the perception of climate change in western Nepal, a study was conducted with the help of 18 KIIs, 185 FGDs, and 27 semi-structured interviews. Selection of respondents from governmental institutions, non-governmental organizations (I/NGOs), educational sectors and research organizations was made from snowball technique on the basis of expertise, professional experience, and knowledge of climate change issues (Jones and Boyd, 2011).

With an objective to assess the climate change experience in the tourism sector, Rayamajhi (2012) conducted a study in Annapurna conservation area of Nepal. The study also identified the relationship between tourism and climate change in the area. For this, a questionnaire survey was conducted with the hotel owners, trekking guides, and travelling visitors. To assess the perception of climate change and tourism. K C and Parajuli (2014) conducted a study in Manaslu Conservation Area of Nepal. The study was done with the help of 5 KIIs, 3 FGDs and 76 questionnaire surveys with local people and secondary data on climate and tourism. Clustered random sampling was done with the help of semi-structured questionnaire in the Nepali language. The field study was conducted in the month of April and May in 2012 in the hilly and mountain region of Nepal.

With an objective to assess the effect of climatic factors on the river system, the study was conducted by Shrestha (2010) in Kathmandu Valley of Nepal. The study was done with the help of climatic data of different stations. To assess the perception of climate change by local people, Dahal (2011) researched on Sano Sirubari VDC of Sindupalchwok District, Nepal. The study was conducted from February to September 2011 with the help of questionnaire surveys, FGDs and KIIS. Pokharel (2011) conducted a study to assess the impacts of climate change in Danchhi VDC of Kathmandu District, Nepal. The research was conducted with the help of questionnaire survey and FGDs. To assess the perception of climate change, Piya *et al.*, (2012) conducted a study in Chepang communities of Nepal. The study was conducted with the help of household survey in mid hill region of Nepal. With an objective to assess annual maximum, minimum and average temperature in the mountain region of Nepal, the study was carried out in mountain slope of south-central Himalayas. There were 13 mountain stations having temperature data lying at an altitude of 1304m to 2566m. Thirty years temperature data from 1980 to 2009 of the stations from DHM were taken for analysis (Kattel and Yao, 2013). With an objective to assess climate change perception of an area, the study was conducted in Sidhi VDC of Chitwan District of central Nepal. For this study, long-term rainfall and temperature data were analysed, and perception of local people towards climate change risk and its impacts on energy was assessed. The primary information was collected from FGDs, KIIs, social resource mapping, energy assessment and seasonal calendar (Shrestha and Shrestha, 2010).

To assess the perception of climate change and its impacts on agriculture, livelihood, ecosystems, and biodiversity, Chaudhary *et al.*, (2011) conducted a study in Kanchenjunga area of Nepal. The study was carried out with the help of field observation, questionnaire surveys, KIIs and FGDs. With an objective to assess the impact of climate change on biological resources and livelihood of people, the study was carried in the northeastern part of Nepal. The study was based on perception analysis of indigenous people in the 5 Village Development Committees of Ilam and Panchthar districts. Primary as well as secondary data on livelihood, culture, religion, and traditions which are closely associated with natural resources were taken for analysis. Secondary information on crop productivity, livelihood and community forestry at VDC, district and national level was taken from published and unpublished sources. Survey and explorative study from a questionnaire survey, FGDs, KIIs and secondary sources were taken for analysis and interpretation. Structured questionnaire surveys of 208 households were taken from 5 different VDCs. Information on crop types, type of biological resources, the trend of rainfall and snowfall pattern, seasonal temperature change, plantation time and adaptation techniques were collected from the household survey (NCDC, 2010).

Results and Discussions

Climate Change Perception

Indigenous people of Kaduna State of Nigeria perceive a change in climate from the historical past due to anthropogenic activities. There was the severe adverse impact on floral and faunal diversity, health, agricultural production and firewood supply (Ishaya and Abaje, 2008). Viner and Agnew (1999) assessed the perception of climate change in international holiday destinations travelled by UK tourists such as the Maldives, The European Alps, Eastern Mediterranean, Southern Spain, Scotland, European Lakes, South and East Africa, Australia, Florida and Brazil with the help of climatic and tourism data. Climate change stress varied according to local conditions and changing environmental systems. Increasing heat stress above 40°C for long time period causes discomfort in many resorts of eastern Mediterranean. Sea level rise was observed in popular tourism destination of small island states of Maldives. Exposure to ultraviolet rays of the sun has increased in Australia due to the decrease in cloud cover. Risk of disease has increased in different parts of the world discouraging tourism. Risk of Malaria is increasing in Spain. Due to a decrease in snowfall, winter tourism of Alps and skiing destinations is adversely affected. Ehmer and Heymann (2008) observed that increase in temperature has negative impacts on Mediterranean region and Austria but slight positive impacts on France, Italy, and Switzerland due to the diversity of tourism structures and reliable snow cover. Most of the countries outside Europe suffer from adverse impacts of climate change with the exception of USA, New Zealand and Canada. It causes more burden on the poor countries where tourism is a main sector of development.

Sem and Moore (2009) observed adverse impacts of climate change on socio-economic aspects of life in small islands and least developed countries. There was a low contribution to global GHG emissions but high risk of existence due to its small size and isolation. It has adverse impacts on tourism, agriculture and other infrastructural works. Simpson *et al.* (2011) observed an increase in temperature in Caribbean tourism destinations in the summer season. Changing climate increased operation cost of tourism entrepreneurs for insurance and temperature regulation. There is increasing cost of pest management, drinking water supply, and irrigation. Climate change would increase the loss from flood, wind and storm up to 50% by 2030. Natural beach and views of the sea would also be destroyed by rising of sea level. About 58% people in Northern Ireland perceived climate change due to anthropogenic activity and natural processes while 4% people did not believe in climate change. Out of these, 57% people observed the impact of climate change on different sectors. About 31% people experienced direct impact of climate change on human being while 48% people do not feel the direct impact on them. The major impacts of climate change were the destruction of nature and wild fauna, heavy flood events, air pollution, frequent severe weather events and adverse impact on public health. About 55% people perceived the impact of climate change on natural environment and wildlife, 54% people perceived the impact of climate change on flooding, 49% people perceived the impact of climate change on atmospheric pollution, 46% people perceived the impact of climate change on severe weather events and 43% people perceived the impact of climate change on public health (DOE, 2009).

Climate Change Perception in Nepal

Analysis of temperature data from 1975 to 2005 in Nepal shows that average temperature was rising by 0.04°C/year which was more than the global average. The average temperature was increasing throughout the country with the exception of few northeast and southern plains. It was observed that temperature of higher altitude and urban areas was increasing at the faster rate than the lower altitude and rural areas (Baidya *et al.*, 2007). The rate of increase in temperature of Kathmandu was 0.05°C/year which was higher than the national average. The seasonal and decadal temperature of Kathmandu was also an increasing trend. The rate of increase in temperature in different seasons was variable with a high rate of increase in the winter. There was a change in extreme temperatures with an increase in hot days and nights and a decrease in cold days and nights. Monsoon rainfall pattern changed from year to year during 1971 to 2005, while the increase in annual average rainfall was 2.08 mm/year. The number of heavy precipitation events has increased while the number of rainy days has decreased when compared to the past (Baidya *et al.*, 2007).

Average annual temperature has increased in all districts with the highest increase in higher elevation. Winter temperature has increased at a higher rate than the summer temperature. The number of hotter days has increased and the number of colder nights has decreased in the hilly and mountainous region while the increased number of cold days and cold nights was perceived in Terai region. Rainfall trend has increased in Palpa and Mustang districts while it has decreased in Ilam, Dolakha and Kailali districts. Rise in temperature, change in rainfall pattern, decrease in rainy days and rainfall amount with high intense rainfall was perceived in these areas. In Jumla, there was the perception of climate change in the form of an increase in temperature, decrease in rainfall and snowfall, drying of water sources, the occurrence of floods and landslides, decrease in productivity of apple, chilly and barley, early flowering, and increase in pests and diseases. Also, fog, cold wave, droughts, and flood were perceived in Terai region (MoSTE, 2015a).

In the Bhaktapur district, there is an increase in temperature and decrease in rainfall in comparison to the past. There was a problem with drinking water supply and water for irrigation due to drying of water sources. In the Godavari area, climate change was perceived in the form of change in rainfall pattern, less rainfall in winter, disturbance in monsoon rainfall, drying of surface wells and reduction in water sources. Increase in temperature, decrease in rainfall, decrease in productivity of vegetables, the occurrence of flood, landslide and droughts and drying of sources of water was observed in the Dhading district. Local residents of the Baglung district perceived that climate change in the form of drying of natural sources of water, unusual flowering and fruiting time of plants, unusual breeding time of cows and buffalos, the occurrence of landslides, floods and droughts and damage of physical infrastructure such as bridges and roads (MoSTE, 2015a).

Climate change has an adverse impact on the livelihood of agriculture and animal husbandry dependent Nepalese farmers. Local farmers experience a lack of water in some places and an excess of water in other places. There is a decrease in agricultural production and an increase in disease and crop pests, human health problems and people's work burden (Macchi *et al.*, 2011). People are experiencing decreasing winter coldness and increasing in hotness in summer. Also, there is decreasing trend of rainfall in rainy season and winter season which is affecting the crop productivity. Price of crops has increased due to the destruction of crops by changing climate and diseases. There is also a decrease in forest resources, firewood, fodder and diversity of wild flora and fauna. Flood events and droughts during rainy season have increased when compared with the past (NCDC, 2010). Observed data of rainfall and AGDP shows that an increase in rainfall amount has a positive impact on AGDP while the increase in temperature has a negative impact on AGDP in Nepal. Increase in rainfall improves the irrigation scheme while the increase in temperature increases evaporation rate. AGDP would increase by Rs. 9.6 million with an increase in one milliliter of rainfall, but AGDP would decrease by Rs. 542 million with the one-degree rise in temperature (Acharya and Bhatta, 2013).

An increased warming trend, rainfall variability, shifts in the production system, change in resource use pattern, and severe impact on agriculture, water resources and health sector was observed in the different region of Nepal (Regmi and Bhandari, 2013). Also, climate change was experienced in the form of an increase in temperature, unfavorable rainfall events, and prevalence of natural disasters such as landslides, floods, droughts and snowmelt (MoSTE, 2015b). There was a reduction in snowfall frequency and amount as compared to the past in mountainous areas of Nepal. Crop productivity and potato yield also decreased due to the death of smaller plants by decreasing insulation capacity of snow. Climate change accelerated food insecurity, biodiversity and natural ecosystem (Rai and Gurung, 2005).

There is a shift in the start of monsoon from June to August and the end of monsoon from September to November. There is also a shift in the dry season from October-November to June-August and snowfall season from November–December to April–May. Snowfall frequency and the amount has decreased in comparison with the past. There is a decrease in production of wheat, barley, millet, and buckwheat in comparison with the past. People are searching crops such as buckwheat, which can grow in dry season and help to earn money. Indigenous knowledge is used to enhance their adaptive capacity and enhance agricultural production. Reduction in crop yield by longer dry spell has forced Dalit girls to leave school and involve in day labour services. The decrease in crop productivity and food insecurity has forced men towards cross-border trade to generate income (Onta and Resurreccion, 2011). In the Banke district of Nepal, the increase in temperature, droughts, and floods is understood in relation to climate change and the adverse impacts it has on rain-fed agriculture, the livelihood of poor and indigenous people and socio-economic sector (GWPNepal, 2014a).

In the Gorkha district, there is an increase in mean monthly temperature from 2001-2006 with the highest variation of 2.1°C in October. There has been a decreasing trend in annual rainfall since the 1980s. The diversity of insects and pests of invasive nature have changed with movement of *Eupatorium odoratum* in higher elevation. Flowering and ripening season of *Cucumis sativus, Luffa cylindrica, Trichosanthes cucumerina, Cucurbita moschata* and other crops have changed. Plantation time of crops has also changed with the development of improved varieties (Lamichhane and Datt, 2009). During an awareness-raising programmes in Gorkha, local people shared their experiences of changes in the climatic pattern. There is increasing, no winter rain with very few high intense rainy days in summer. The number of hot and humid days are increasing in summer while cold days are decreasing (MoPE, 2016b). In Manaslu Conservation Area of Gorkha, there is an increase in annual maximum temperature and mean temperature by 0.10 and 0.02°C/year, decrease in annual minimum temperature by 0.06°C/year and increase in average annual rainfall by 3.19mm/year. About 93% of the sampled population perceive a change in climate, 36% perceive the most significant impact of climate change on rainfall and water supply and 82% perceive unfavourable weather change. Rising temperature, rapid melting of snow, increase in rainfall amount in the lower belt, a decrease in snowfall in the higher belt and unfavourable weather change phenomena are observed in the area (K C and Parajuli, 2014).

Rise in temperature, high intense rainfall, decrease in rainfall volume, drying of water sources, landslides, and spread of mosquitoes, pests and diseases are observed in Bandipur. There is a decrease in volume of water in the stream in pre-monsoon season which troubles the communities. The rise in temperature and change in rainfall pattern affects water quantity and causes water scarcity (MoSTE, 2015a). Increased frequency of landslides and floods destroys irrigation system, water supply, and water mills. Local indigenous knowledge and skills are helpful to adapt to the impacts of climate change and manage water and forest. Forest ecosystems are disturbed by increased fragmentation, change in species diversity and growing season, decrease in productivity and increase in forest fires and floods (MoSTE, 2015a). In a study conducted by Bhusal (2009) in Lumle and Ghandruk VDC of Kaski district, 90% of respondents' experience increase in temperature and 97% perceived changing rainfall in the last 10 years. About half of the people perceived late start of the monsoon, increase in a dry spell, unexpected hail storm, high wind flow trend, and drying of water sources. Only 2% of respondents perceive a decrease in temperature while 3% perceive constant rainfall patterns. About 72% perceive occurrence of drought and unfavourable rainfall patterns in past few years.

Lama (2010) perceived warm and dry winter and a decrease in winter snowfall in Lower Mustang. There is a positive impact on tourism and negative impact on the natural beauty and water availability. Due to drying up of existing springs, there are potable water problems and a need for new sources. Droughts and forest fire increase due to increase in temperature. Local residents of Mustang perceive high-intensity rainfall with an increase in summer precipitation. Winter temperature increased as compared to the past resulting in less cold and less frosty environment. Rainfall pattern changed with the occurrence of heavy rain with a long dry spell. There is shifting of winter snowfall from December to February affecting plant growth due to low moisture. Due to the early melting of snow in March, greenery has reduced resulting in low fodder production. There is a shortage of drinking water due to low flow of water in the streams and springs. Irregular flow in spring and stream results in migration of people of northern Mustang. There is heavy rainfall in the southern part instead of snowfall in the winter season. It damages houses which were made of mud. Climatic hazards like floods, hail storms, wind storms and avalanches are more frequent as compared to that of past. Apple fruits, green plants, and other crops are destroyed by hailstorms. Increase in temperature creates dry and warm weather which is more comfortable for old people and tourists. Unfavourable rainfall pattern disturbs the whole ecosystem by reducing the soil moisture in the dry season and increasing soil erosion in the wet season (CEPAD, 2011).

Sixty nine (69) % of hotel owners and guides perceive a change in tourism season due to climate change in Annapurna conservation area of Nepal. Forty-nine (49) % of the respondents perceive a relationship between climate change and tourism while 34% feel no any relationship and 17% are not sure about the relationship. Eighty-three (83) % of hotel owners and guides perceive that climate change is occurring which could affect the number of tourist flow in the area. Rising temperature has not caused any discomfort to the tourists, but unfavourable rainfall hampers hotel owners business. The rise in temperature causes decrease in snow cover and melting of the glacier (Rayamajhi, 2012). Samzong, Dhey and Yara village of Upper Mustang are in the process of retreat due to water scarcity, lack of grazing land and flood risk (MoSTE, 2015a).

The temperature of the south-central Himalayas in Nepal is increasing at the rate of 0.40 to 0.91°C/decade. On average, the temperature of 13 mountain stations is increasing at the rate of 0.38°C/decade. There is the highest increase in maximum temperature by 0.65°C/decade and lowest increase in minimum temperature by 0.11°C/decade. Increase in temperature of Khumaltar stands at 0.53°C/decade, and Godavari is 0.46°C/decade (Kattel and Yao, 2013). Increase in temperature, changes in rainfall trend, increase in flood events and long dry spell are observed in western Nepal. It has adverse impacts on agricultural production and working habit of people (Jones and Boyd, 2011). People of Nepal experience unexpected floods, unstable rainfall, landslides, soil erosion, hot air, dry spells, thundershowers, the spread of mosquitoes and insects. People of Kathmandu perceive an increase in mosquito bites, snakebite, malaria, dengue, encephalitis, and filariasis. People of rural areas perceive a loss of water, the untimely flowering of *Rhododendron arboretum* and unusual ripening of Kafal fruit (Sharma, 2009).

Climate change has adverse effects on the life, food sources, water resource and disaster management to the people of Khokana and other parts of Nepal (GWPNepal, 2014b). In the Kathmandu valley, 75% of annual rainfall occurs in the monsoon season. There is sporadic rainfall from October to May with more dryness. Seven stations out of nine have a decreasing trend, while Khumaltar Station has increasing trend and Bhaktapur station has a constant hydrograph. There is decreasing trend of rainfall resulting in low river discharge and groundwater recharge (Shrestha, 2010). About 37% of Chepang people of the mid-hills region of Nepal perceive a decrease in the total rainfall, 23% feel unperturbed rainfall, 21% experienced late rainfall, and 13-16.7% observe no changes in rainfall. About 48% people perceive an increase in summer temperature, 10% feel cooler summer, 22% perceived cooler winter, 23% feel warmer winter, and 39% do not perceive any changes in temperature (Piya *et al.*, 2012).

In Sano Sirubari VDC of Sindhupalchwok District, warmer days and less frequent winter days are observed. Rainfall trend has changed, thereby decreasing drinking water sources and irrigation water. Agricultural production of rice, maize, and millet have reduced with an increase in diversity of invasive species, pests, and insects. The decrease in production of fodder has reduced the growth of livestock which in turn reduces the income generated from it. It has compelled people to leave the traditional occupation and search for other alternatives. Human health problems such as skin diseases, menstruation cycle, eye diseases, diseases of uterus and itching problem have increased due to climate change. The spread of mosquitoes has also increased. Flowering season of *Shorea robusta*, Amala and maize have changed. Germination, harvesting, and maturation of different agricultural crops have changed (Dahal, 2011). People of Danchhi VDC experience different types of hazards such as flood and dry spell due to climate change. Also, people perceive unfavourable rainfall in different seasons and snowfall and hailstone in winter (Pokharel, 2011).

During Vulnerability analysis of Jugedi Khola watershed in Chitwan district, Nepal, 99% of the respondents reported landslides and floods as major hazards in the area induced due to the erratic rainfall. Climate change also causes soil erosion, lack of irrigation facility and seasonal droughts affecting the well-being of the communities (Gurung and Bhandari, 2009). Local people of Chitwan district of Nepal perceive climate change in the form of an increase in temperature, change in rainfall pattern, increased intensity of dry spells, the emergence of invasive species, pests, diseases and adverse effects on human health. People also perceive drying of water sources, thick foggy mornings, alteration of flowering and fruiting of invasive species such as *Mikenia macarantha, Eupatorium adenophora* and *Lantana camera*, a decrease in fish diversity, modification of river channels and sedimentation. Hot summers, mild winters, late onset and faster ending of rainfall and spread of mosquitoes are also observed in Chitwan district (Shrestha and Shrestha, 2010).

Chaudhary *et al.*, (2011) observed that people in the Kanchenjunga region are aware of climate change and its impacts on ecosystems, biodiversity, weather, and agriculture. More than 84% respondents perceive an increase in temperature and about 78.6% perceive late onset of monsoon and summer season in past 10 years. About 40% respondents perceive more frequent occurrence of a dry spell and 68.8% respondents observe mosquitoes in lower hills. More than 75% respondents observe drying up of water sources, and more than 60% respondents felt less snowfall in the Himalayas. About 56.8% respondents perceive faster ripening of vegetables and crops.

Conclusion and the Way Forward

Climate change is a global as well as national issue which affects a biological, physical, social and economic sector of the environment. It is observed in the form of an increase in temperature and change in the usual trend of rainfall in Nepal. With an objective to assess climate change perception in Nepal, this chapter was prepared with the help of literature review of different research reports and publications. Researchers have applied different methods for assessing climate change perception. They focused on a questionnaire survey, field observations, key informant interviews, focus group discussions, climatic data collection and modelling for climate change perception. People perceive a rise in summer and winter temperature, change in rainfall trend in lower altitude, decrease in snowfall in higher altitude and high melting of the glacier as a result of climate change. Climate change causes a decrease in agricultural production, shrinking water resources, adverse health impacts and changed livelihood. To reduce the impact of climate change, there is a need for accurate climatic data collection and forecasting, effective early warning system, vulnerability assessment, and awareness.

It is difficult to study impacts of climate change in Nepal due to the lack of long interval instrumental climatic data (NCDC, 2010). Accurate data collection and forecasting, effective early warning system, adequate mapping of hazard and vulnerability, and awareness generation would minimize risks of climate change. Use of glacial water in hydropower, irrigation and storage reservoir would help to minimize climatic risk (Alam and Regmi, 2004). To address the adverse impacts of climate change, an integrated and coordinated approach to improving our understanding of climate change science, impacts, vulnerability, adaptation, and mitigation should be prioritized. There is a need for awareness, advocacy and negotiations to minimize climate change impacts and increase resilience. For enhancing climate change adaptation options, economic, social and ecological dimensions of sustainable development should be focused (Parajuli, 2010). Impacts and adaptation strategies towards climate change need to be identified with the proper planning process. Shifting of habit towards low carbon economy could minimize adverse impacts of climate change. Indigenous knowledge needs to be documented, verified and replicated in a different region of Nepal (NCDC, 2010). Community forest management in Nepal needs to be enhanced which could help to absorb atmospheric carbon dioxide and mitigate global climate change to some extent (K C *et al.*, 2016; K C *et al.*, 2014).

References

Acharya, S. P., & Bhatta, G. R., 2013, Impact of Climate Change on Agricultural Growth in Nepal Nepal Rastra Bank Working Paper 15. Kathmandu, Nepal: Nepal Rastra Bank.

ADB, 2015, The Asian Development Bank And The Climate Investment Funds Country Fact Sheets. Manila, Philippines: ADB Climate Change and Disaster Risk Management Division.

Agrawala, S., Raksakulthai, V., Aalst, M. V., Larsen, P., Smith, J., & Reynolds, J., 2003, Development and Climate Change in Nepal: Focus on Water Resources and Hydropower. Paris, France: Organisation for Economic Cooperation and Development.

Alam, M., & Regmi, B. R., 2004, Adverse Impacts of Climate Change on Development of Nepal: Integrating Adaptation into Policies and Activities. Dhaka, Bangladesh: Bangladesh Centre for Advanced Studies.

Baidya, S. K., Regmi, R. K., & Shrestha, M. L., 2007, Climate Profile and Observed Climate Change and Climate Variability in Nepal. Kathmandu, Nepal: Department of Hydrology and Meteorology.

Bhusal, Y.R., 2009, Local Peoples' Perceptions on Climate Change, Its Impacts and Adaptation Measures in Mid-Mountain Region of Nepal (A Case study from Kaski District). B.Sc. Forestry, Tribhuvan University, Pokhara, Nepal.

CBS, 2012, National Population and Housing Census 2011 (1). Kathmandu, Nepal: Central Bureau of Statistics.

CCNN, 2011, Governance of Climate Change Adaptation Finance in Nepal. Kathmandu, Nepal: Climate Change Network Nepal.

CEN, 2012, Information note for COP18 to the UNFCCC and CMP8 to the Kyoto Protocol. Kathmandu: Clean Energy Nepal.

CEPAD, 2011, District Climate and Energy Plan for Mustang District. Lalitpur, Nepal: CEPAD Consultants (P) Limited.

Chaudhary, P., Rai, S., Wangdi, S., Mao, A., Rehman, N., Chettri, P., & Bawa, K.S., 2011, Consistency of Local Perceptions of Climate Change in the Kangchenjunga Himalaya Landscape. CURRENT SCIENCE, 101(4), 10.

Dahal, D.S., 2011, Impact of Climate Change on Livelihood and Biodiversity in Rural Communities, A Case Study of SiddhiGanesh and Nepane Community Forestry User Groups of Sindhupalchwok District of Nepal. Master, Tribhuvan University, Kirtipur, Kathmandu.

Dahal, N., Ojha, H., Baral, J., Branney, P., & Subedi, R., 2009, Impact of Climate Change on Forests and Livelihoods: Issues and Options for Nepal. Kathmandu, Nepal: Livelihoods and Forestry Programme.

DOE, 2009, Public perception on climate change in Northern Ireland 2009 Northern Ireland: North Ireland Statistics and Research Agency North Ireland Environment Agency.

Ehmer, P., & Heymann, E. 2008, Climate Change and Tourism, Where Will the Journey Lead? In T. Just (Ed.). Frankfurt, Germany: Deutsche Bank AG, DB Research.

FAO, 2010, Implications of Climate Change for Agriculture and Food Security and Adaptation Priorities in Nepal. Rome, Italy: Food and Agriculture Organization of the United Nations.

GoN, 2011, Climate Change Policy, 2011. Kathmandu, Nepal: Government of Nepal.

GoN, 2015, Nepal Disaster Report 2015. Kathmandu, Nepal: The Government of Nepal, Ministry of Home Affairs (MoHA) and Disaster Preparedness Network-Nepal (DPNet-Nepal).

Gurung, G. B., & Bhandari, D. (2009). Integrated Approach to Climate Change Adaptation. Journal of Forest and Livelihood 8(1), 90-99.

GWP Nepal, 2014a, Climate Vulnerability and Gap Assessment Report on Flood and Drought (Lower Rapti River Basin Case Study) Kathmandu, Nepal: GWP Nepal/Jalsrot Vikas Sanstha (JVS).

GWP Nepal, 2014b, Traditional Climate Change Adaptation Practices by farmers in Nepal Kathmandu, Nepal: Global Water Partnership Nepal/ Jalsrot Vikas Sanstha (JVS).

IPCC, 2007, Climate Change 2007: The Physical Science Basis Summary for Policymakers, Contribution of working Group 1 (WG1) to the Fourth Assessment Report of the IPCC (Pre- Publication ed.).

Ishaya, S., & Abaje, I. B., 2008, Indigenous People's Perception on Climate Change and Adaptation Strategies in Jema'a Local Government Area of Kaduna State, Nigeria Journal of Geography and Regional Planning, 1(8), 6.

Jones, L., & Boyd, E., 2011, Exploring social barriers to adaptation: Insights from Western Nepal. Global Environmental Change, 21, 1262–1274. doi: 10.1016/j.gloenvcha. 2011.06.002

Kattel, D. B., & Yao, T., 2013, Recent temperature trends at mountain stations on the southern slope of the central Himalayas. J. Earth Syst. Sci. , 122(1), 215–227.

K C, Anup, 2017, Climate Change Communication in Nepal. In W. Leal Filho et al. (eds.), Handbook of Climate Change Communication: Vol. 2, Climate Change Management, 21-35, Springer International Publishing AG

K C, Anup & Ghimire, A., 2015, High-Altitude Plants in Era of Climate Change: A Case of Nepal Himalayas. In M. Ozturk, K. R. Hakeem, I. Faridah-Hanum & R. Efe (Eds.), Climate Change Impacts on High-Altitude Ecosystems (1), 177-189. Switzerland: Springer International Publishing.

K C, Anup & Thapa Parajuli, R. B., 2014, Climate Change and Its Impact on Tourism in the Manaslu Conservation Area, Nepal. Tourism Planning & Development, 13. doi: 10.1080/21568316.2014.933122

K C, Anup, 2015, Climate Change and its impact on hydropower development in Nepal. Vidhyut, 26(1), 25-29.

K C, Anup, Bhandari, G., Joshi, G. R., & Aryal, S., 2013, Climate Change Mitigation Potential from Carbon Sequestration of Community Forest in Mid Hill Region of Nepal International Journal of Environmental Protection, 3(7), 33-40.

K C, Anup, Joshi, G. R., & Aryal, S., 2014, Opportunity Cost, Willingness to Pay and Cost Benefit Analysis of a Community Forest of Nepal. International Journal of Environment, 3(2), 108-124.

K C, Anup, Manandhar, R., Paudel, R., & Ghimire, S., 2016, Increase of Forest Carbon Biomass due to Community Forestry Management in Nepal. Journal of Forestry Research.

Lama, A. K., 2010, Vulnerability of Nature Based Tourism to Climate Change: Stakeholders' Perceptions of and Response to Climate Change in the lower Mustang Region of the Annapurna Conservation Area. Master of Parks, Recreation and TourismManagement Lincoln University, Christchurch, New Zealand.

Lamichhane, Babu R., & Datt, A. K., 2009, Changing Climate in a Mountain Sub-watershed in Nepal Journal of Forest and Livelihood, 8(1).

Macchi, M., Gurung, A. M., Hoermann, B., & Choudhary, D. 2011, Climate Variability and Change in the Himalayas: Community Perceptions and Responses. Kathmandu, Nepal: International Centre for Integrated Mountain Development.

Maharjan, M., 2014, Climate Change Adaptation Projects and Major Activities on Climate Financing in Nepal. Kathmandu, Nepal: Global Water Partnership (GWP) Nepal/ Jalsrot Viksa Sanstha (JVS).

MoE., 2011, National Framework on Local Adaptation Plans for Action. Babarmahal, Kathmandu, Nepal: Government of Nepal Ministry of Environment.

MoPE. 2016a, Climate Change Division Retrieved September 14, 2016, from http://www.mope.gov.np/index.php#

MoPE. (2016b). eBulletin. from Ministry of Population and Environment, Government of Nepal, Kathmandu, Nepal

MoPE. (2016c). Nepal Climate Change Support Programme, Ministry of Population and Environment, Government of Nepal, Kathmandu

MoSTE. (2015a). Indigenous and Local Climate Change Adaptation Practices in Nepal. Kathmandu: Ministry of Science Technology and Environment, Government of Nepal.

MoSTE. (2015b). Indigenous and Local Knowledge and Practices for Climate Resilience in Nepal. Kathmandu, Nepal: Mainstreaming Climate Change Risk Management in Development, Ministry of Science, Technology and Environment (MoSTE).

MoSTE. (2015c). National Adaptation Plan Formulation Process Kathmandu, Nepal: Government of Nepal Ministry of Science, Technology and Environment Climate Change Management Division.

MoSTE. (2015d). Nepal Climate Change Program (CCP) Program Progress Report January 1– December 31, 2014 Kathmandu, Nepal: Ministry of Science, Technology and Environment.

NCDC. (2010). Understanding Indigenous People's Perception on Climate Change Impact on Floral and Faunal Species in the Kanchenjunga Singalila Complex, Eastern Nepal. Dhdobidhara, Ilam, Nepal: Namsaling Community Development Centre.

Onta, N., & Resurreccion, B. P. (2011). The Role of Gender and Caste in Climate Adaptation Strategies in Nepal, Emerging Change and Persistent Inequalities in the Far-Western Region. Mountain Research and Development, 31(4), 351-356.

Parajuli, R. (2010). Climate Change and Local Impacts-Synergy between Adaptation and Renewable Energy Development for Nepal. Kathmandu, Nepal: Alternative Energy Promotion Centre.

Piya, L., Maharjan, K. L., & Joshi, N. P. (2012). Perceptions and Realities of Climate Change among the Chepang Communities in Rural Mid-Hills of Nepal. Journal of Contemporary India Studies: Space and Society, Hiroshima University, 2, 16.

Pokharel, P. (2011). Documentation of Climate Change Impacts, Community Perceptions, and Adaptation Practices in a Rice Farming Community in Nepal All Nepal Peasants' Federation (ANPFa).

Rai, S. C., & Gurung, A. (2005). Raising Awareness of the Impacts of Climate Change Initial Steps in Shaping Policy in Nepal. Mountain Research and Development, 25(4), 316-320.

Rayamajhi, S. (2012). Linkage between Tourism and Climate Change: A Study of the Perceptions of Stakeholders along the Annapurna Trekking Trail. Nepal Tourism and Development Review 2(1), 60-86.

Regmi, B. R., & Bhandari, D. (2013). Climate Change Adaptation in Nepal: Exploring Ways to Overcome the Barriers Journal of Forest and Livelihood, 11(1), 43-61.

Regmi, B. R., Morcrette, A., Paudyal, A., Bastakoti, R., & Pradhan, S. (2010). Participatory Tools and Techniques for Assessing Climate Change Impacts and Exploring Adaptation Options, A Community Based Tool Kit for Practitioners. Kathmandu, Nepal: Livelihoods and Forestry Programme.

Sem, G., & Moore, R. (2009). The Impact of Climate Change on the Development Prospects of the Least Developed Countries and Small Island Developing States: Office of the High Representative for the Least Developed Countries, Landlocked Developing Countries and Small Island Developing States (UN-OHRLLS).

Sharma, K. P. (2009). Climate Change Trends and Impacts on Livelihood of People Kathmandu, Nepal: Jalsrot Vikas Sanstha/Nepal Water Partnership.

Shivakoti, B. R., Lopez-Casero, F., Kataoka, Y., & Shrestha, S. (2015). Climate change, changing rainfall and increasing water scarcity: An integrated approach for planning adaptation and building resilience of smallholder subsistence livelihoods in Nepal IGES Research Report 2014-07. Hayama, Japan: Institute for Global Environmental Strategies (IGES).

Shrestha, A. B., Wake, C. P., Mayewski, P. A., & Dibb, J. E. (1999). Maximum Temperature Trends in the Himalaya and its Vicinity: An Analysis Based on Temperature Records from Nepal for the period 1971-94. Journal of Climate, 12, 15.

Shrestha, P. (2010). Climate Change Impact on River Dynamics of the Bagmati Basin, Kathmandu Nepal. Kathmandu, Nepal: National Adaptation Program of Action (NAPA) to Climate Change Project, Government of Nepal.

Shrestha, S., & Shrestha, A. (2010). Gender Perspective: Integrating Energy Resource Use into Climate Change Adaptation. Kathmandu, Nepal: National Adaptation Program of Action (NAPA) Project office, Ministry of Environment, Singha Durbar.

Simpson, M., Scott, D., & Trotz, U. (2011). Climate Change's Impact on the Caribbean's Ability to Sustain Tourism, Natural Assets, and Livelihoods: Inter-American Development Bank.

Viner, D., & Agnew, M. (1999). Climate Change and Its Impact on Tourism. Norwich UK. NR47T: Climate Research Unit (CRU), University of East Anglia.

Chapter Six

STAKEHOLDERS IN CLIMATE CHANGE COMMUNICATION IN THE NORTHWEST REGION OF CAMEROON

Suiven John Paul Tume, *Green Care Association, Cameroon*

Moye Eric Kongnso, *University of Dschang, Cameroon*

Mbu Dora Nyukighan, *Federal University of Lokoja, Nigeria*

Nyuykongadzem Emma Dindze,
University of Bamenda, Cameroon

Njodzeka Gilbert Njodzeka, *Green Care Association, Cameroon*

Abstract

Climate change communication is relatively new in Cameroon. It has become prominent because of dramatic environmental changes that threaten livelihoods. This paper, therefore, sets out to examine the role of different stakeholders in communicating climate change. To ascertain this, 29 media outlets composed of 26 radio stations and 3 newspapers were sampled to analyse their programs on climate and environmental issues throughout the Northwest Region of Cameroon. In spite of their efforts to sensitize the public on the changing climate, the media fails to communicate the basics like a daily weather forecast. However, 20 media houses (68.69%) broadcast climate change-related programs, and 20 (68.69%) have other environmental related programmes. All the media houses reported that the public is receptive to climate and environmental issues that they broadcast. Only 12 of the media outlets use the social media (41.37%) to disseminate climate concerns. A number of Non-Governmental Organisations and Common Initiative Groups were also sampled to know how relevant their visions and missions fit into contemporary climate change and environmental communication. The public is actively engaged in climate change communication by media houses through interactive radio programs, invitation of resource persons to deliberate on climate issues, field investigations and reporting of evidence and vulnerabilities of the changing climate and seeking experts to address such concerns. Some of the radio programs are aired by Civil Society Organisations (CSOs) under the canopy of the Network of Environmental Stakeholders Cameroon (NEST-Cam). In spite of the fact that it is gaining ground in the Northwest Region of Cameroon, there is a need for more interactive radio programs. Journalists need more capacity on climate/environmental reporting.

Key words: Adaptation • Communication • Media • Stakeholders

Introduction

Climate change communication has lately become conspicuous in society from movies to grassroots movements (Nerlich *et al.*, 2009; Ward and Menezes, 2008). It is a global problem with widespread impacts. It is crucial that climate change messages are communicated effectively by different stakeholders including the media, Non-Governmental Organisations, socio-cultural groups, municipal councils and government institutions (European Spatial Planning: Adapting to Climate Events-ESPACE, 2007). Mitigation is concerned with the causes of global warming and calls for the reduction of greenhouse gas emissions. Adaptation, on the other hand, is concerned with the impacts on society, economy, and environment, thereby, promoting activities to reduce vulnerability to extreme weather events (floods and droughts) and other long-term changes in climate (ESPACE, 2007; IPCC, 2014). Climate change communication and its impact on the general public have proliferated in communication and related disciplines since the 1990s (Nerlich *et al.*, 2009). In spite of this, society remains vulnerable to climate variability. This raises questions about the effectiveness of communication efforts and the ability of audiences to implement adaptation in response to these communications (Moser, 2010; Monroe *et al.*, 2015). This concerns persuasiveness of the messages, the structure of society and considerations of the extent to which the public is receptive to make an effective change (Nerlich *et al.*, 2009; IPCCC, 2014; UNFCCC, 2006). The combination of natural and human-induced factors accelerate climate change, and consequences abound (interruptions in agriculture, surges in the frequency of droughts and floods) (Filho, 2009). Since human-induced climate change first emerged on the public agenda in the mid-1980s, public communication of climate change and more recently the question of how to communicate it most efficiently have witnessed a sharp rise (Moser, 2010; Monroe *et al.*, 2015; Nerlich *et al.*, 2009).

Materials and Methods

The Northwest Region is the third most populated region in Cameroon. It is located at 6°20'N-6°33'N and 10°30'E-10°50'E coordinates, with a surface area of about 17,300 km^2. The capital city is Bamenda. The region saw an increase in its population from about 1.2 million in 1987 to an estimate of 1.8 million in 2010 (Ngonga and Tume, 2015). The population density of 99.12 pers./km^2 is higher than the national average of 22.6 pers./km^2. The urban growth rate is 7.95 %, higher than the national average of 5.6 %, while the rural growth rate is approximated at 1.16 % (Lambi, 2001). The region was created in 1972 with five administrative Divisions-Bui, Donga-Mantung, Menchum, Mezam and Momo. Today, it has seven divisions. The additions are Boyo-carved out of the Menchum and Ngoketunjia, split off from Mezam (Fig. 6.1). Each division is further separated into sub-divisions. The basic unit of local government is the council. There are thirty-two councils in the region.

Fig. 6.1: Location of the Northwest Region of Cameroon

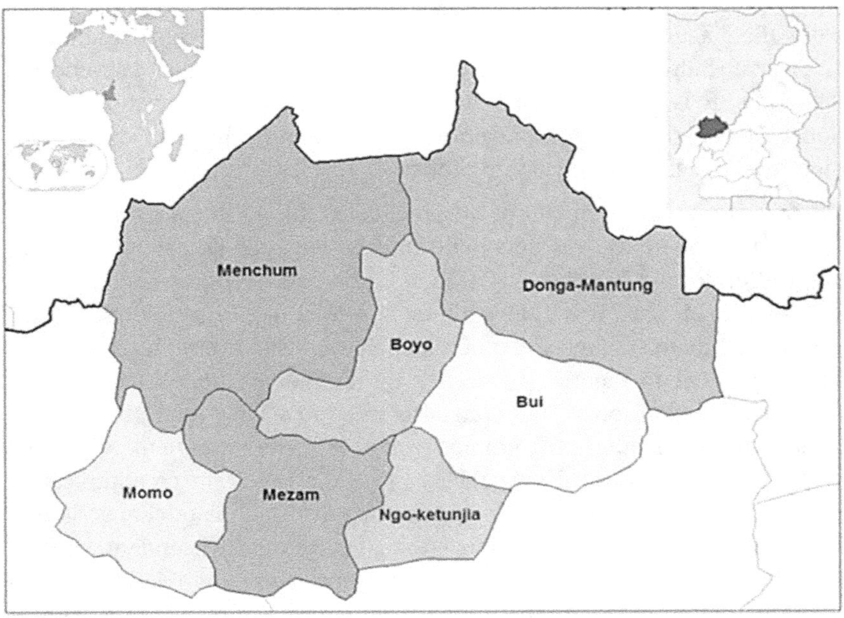

(Source: Wikipedia, accessed 20/11/16)

The Northwest Region of Cameroon is a part of the mountainous grassland savannah zone of Central and West Africa with the *'Aw'* climate type (tropical grassland). The high altitude farming zones with volcanic soils are rich in organic matter with an annual average rainfall of about 2,500mm. The area has great potentials for intensive agriculture (Molua and Lambi, 2007). The government, Non-governmental Organisations (NGOs), the media and socio-cultural groups have taken up initiatives to sensitise the population in the fight against climate change. For an ordinary farmer who lives in a rural area and for most people in town, the issue of climate change is not a strange phenomenon. People relate their visible experiences of changing weather patterns to global warming. This makes them more engaged with the media in discussions that enable them to make sense of the perceived and visible changes. The fact that people speak about global warming all the time means that there is a new message circulating in a society that changes the way people perceive the world and their relationship with the environment.

Cash flows from international donor organisations as many NGOs in the Northwest Region of Cameroon have shifted their policies to work on climate change-related issues. Part of these projects is sensitisation and awareness campaigns in which people are discouraged from burning their land and cutting down trees (de Wit, 2011). Moreover, at the grassroots level, planting trees has become an imperative in the call for thinking globally and acting locally. However, NGOs do not operate in isolation but rather jointly implement their activities. In November 2009, an association called Cameroon Traditional Rulers Against Climate Change (CAMTRACC) was launched in Bamenda (de Witt, 2011). Inspired by the FAO of the United Nations, traditional rulers of the Northwest Region united to form a solid front at the grassroots level to fight against climate change.

In order to ascertain the role of stakeholders in climate change communication, ethical clearance was obtained from the Regional Delegation of Communication for the Northwest Region of Cameroon to survey media outlets. The instrument used was a questionnaire. A total of 29 media outlets composed of 26 radio stations and 3 newspapers were surveyed. The media outlets were asked to identify themselves by name, type (radio, television or print) and location town. They were also asked to specify their programs on climate as well as other programmes related to the environment. Again, the media outlets were asked how they engage the public in communicating climate change issues and if the public receptive to climate change issues. Since climate change is linked to several aspects of the environment, another question was directed towards international environmental-related days. An additional question was on the use of social media in disseminating climate change information.

The last question was on manifestations of climate change around the local environment. The responses were expressed as percentages and presented in tables. Some of the radio stations are not fully aware of all international days related to climate and the environment. This study, therefore, is a sensitisation and outreach activity to raise awareness of climate change and environmental issues. A copy of the questionnaire was retained at each media house to keep them alert of all international days so that they can conduct more investigative journalism for objective reporting and raise more public awareness. A cross-section of NGOs and Common Initiative Groups (CIGs) were also sampled to know how their visions and missions fit in contemporary climate change and environmental communication in the Northwest Region of Cameroon. Some of these Civil Society Organisations (CSOs) are grouped under the canopy of the Network of Environmental Stakeholders Cameroon (NEST-Cam).

Drivers of Climate Change in the Northwest Region of Cameroon

Human activity has caused a variety of changes in different forcing agents in the atmosphere and land surface. A large number of greenhouse gases have increased over time from anthropogenic origin. Atmospheric aerosols have diverse and complex influences on the climate. Human activity has modified the land cover and changed the surface albedo. Some of the gases and aerosols are directly emitted to the atmosphere whereas others are secondary products from chemical reactions of emitted species. The lifetimes of these different forcing agents vary substantially (Myhre *et al.*, 2013). Anthropogenic land cover change has a direct impact on the Earth radiation budget through a change in the surface albedo. It also impacts the climate through modifications in the surface roughness, latent heat flux, and river runoff.

The landscape of the Northwest Region of Cameroon is dominated by remnants of gallery forests under the threat of overgrazing and unsustainable agricultural practices. This stunted highland vegetation is very resistant to prolonged droughts. Slash and burn farming method leads to the destruction of soil organic matter. It is caused by bushfires during the dry season. Bushfires can cause severe consequences for the environment including loss of vegetation and wildlife. These fires also cause changes to the atmosphere such as increased levels of CO_2 in the air through large volumes of smoke and ash and localised change in weather. The impact of bushfires on the people of the region is devastating. Some of the effects are lasting health problems, poverty resulting from loss of livelihood, rural exodus, social displacements in families, intra-and inter-community conflicts from competition over scarce natural resources and many other negative effects such as insufficient pasture for cattle. Cattle grazing is not only an unfriendly environmental activity in the Northwest Region of Cameroon. It triggers more pressure on stunted vegetation. Animals also destroy food crops, thereby intensifying farmer-grazier conflicts. This makes the Northwest Region of Cameroon highly vulnerable to climate change like most areas in Sub-Sahara Africa.

Climate change-related impacts in the Northwest Region of Cameroon include temperature fluctuations from year to year, desiccation of natural habitats, more frequent droughts and floods. Such changes have a negative impact on agricultural production and food security. Similarly, the sustainability of some rural infrastructure would also be negatively affected by climate change. The health of people in the area could be worsened by climate variability (drying up of water resources and flooding are closely related to an increase in water-borne diseases such as cholera).

Another driver of climate change in the Northwest Region of Cameroon is the *'water vampires'*–eucalyptus trees which many consider as *'environmental terrorists'*. Eucalyptus plantations have replaced the indigenous trees. They are used for fuelwood and electric and telecom transmission poles in Cameroon and beyond. Following the slumping of coffee prices in the 1980s, coffee plantations were cut down in favour of eucalyptus. The eucalyptus also provides revenue for plantation owners and municipal councils to the detriment of water resources and the environment. Other extremes of climate change in the Northwest Region are urban flooding in Bamenda, landslides, mudflows, and rockfalls emanating from heavy rainfall. Some of these local realities have been the erratic rainfall, especially the onset of rain at the beginning of the wet season and prolonged dry seasons. These variations in the climate have greatly disrupted the agricultural calendar. These climatic variations also have negative impacts with multiplier effects like drying up of some annual springs, frequent droughts during the rainy season, increasing local temperatures, poor harvests of staple crops like maize, beans and Irish potatoes, incessant water crisis throughout the year, reduction in volumes of rivers, springs and other water bodies. These alterations constitute the climate change messages that these media houses broadcast.

Since 2012, Cameroon has a National Climate Change Action Plan (NCCAP) aimed at building the capacity of socio-economic actors to adjust to climate change (UNDP, 2009; African Development Fund, 2013; Egan, 2013). A series of adaptation measures include:

- assessment of risk and vulnerability studies to identify the best intervention strategies; sensitisation and training of traditional rulers (Fons, Chiefs, Aldors), mayors of municipal councils and farmers on environmental protection and climate change adaptation; knowledge building on climate change adaptation by training many stakeholders; promotion of improved stoves to ease pressure on wood resources; increase in the proportion of reforestation of watersheds through analogue forestry; rangeland improvement activities to reduce bushfires and consequently scale down greenhouse gas emissions.

Stakeholders in Climate Change Communication

The Fifth Assessment Report (AR5) of the Intergovernmental Panel on Climate Change (IPCC) is widely recognised by academics, journalists, communications experts, governments and civil society organizations to have marked strides in the approach the IPCC communicates its agenda (IPCC, 2016). The communication of AR5 saw both greater professionalism at all stages of the process and greater breadth and diversity in the subsequent outreach accomplishments. The results of this can best be seen in the Paris Agreement of the United Nations Framework Convention on Climate Change (UNFCCC) reached at the Conference of Parties in December 2015 (COP-21). This agreement is based on assessments that the IPCC communicated to negotiators through the Structured Expert Dialogue and UNFCCC meetings. The IPCC has also experienced growing calls from policymakers and other users to do more with its communications. Some improvements to IPCC communications came not from the communications team but from the authors of AR5–the use of headlines statements in the Working Groups I, II and III contribution to AR5 and the Synthesis Reports. Other improvements for AR5 included:

- Responding to media questions before completion of the reports; Media workshops to clarify the workings of the IPCC and how it produces assessments; Making IPCC communications more professional by working with external communications specialists; Media training for bureau members and authors; Systematic planning of interviews with a range of authors, both face-to-face and remotely; Arranging facilities for broadcasters; Production of scientifically rigorous but compelling videos, overseen by the working group co-chairs and IPCC Chair; Ambitious programme of outreach activities all over the world; Cooperation with third parties producing versions of the report targeting specific sectors in specific regions; Use of social media to publicise IPCC findings and outreach activities.

This gives the IPCC a strong foundation to build on for its future climate change communications. In line with the IPCC, Change Communications, a not-for-profit, and apolitical organisation trained twenty community radio journalists from ten stations in the Northwest Region of Cameroon on the basic concepts of climate change and techniques of reporting it in March 2016. At the end of their training which held in Bamenda, they were expected to produce carefully tailored programs and professionally sound news reports on climate change issues in their respective localities. There is evidence that the information and messages contained in the programs produced so far are timely and useful for the public. Media houses in the Northwest Region of Cameroon produce a variety of climate-focused programs (71.42 %) and other environmental-related programs (67.85 %) (Table 6.1).

Table 6.1: Climate-Environmental Focused Programs in Northwest Cameroon

Item	Responses	
	% Yes	% No
Daily Weather Forecast	03.44	96.56
Climate-focused programs	68.96	31.04
Other environmental-related programs	68.96	31.04
Public perception of climate change messages	100.0	00.00
Use of social media to communicate climate change	41.37	58.63

Source: Fieldwork, May-August 2016

The public is receptive to these programs (Ward and Menezes, 2008), though there is limited use of social media to engage the youth in climate change communication fully. Other stakeholders involved in the training of journalists on climate and environmental reporting in Cameroon include Media Synergy for the Promotion of Biodiversity Conservation and Community Development against Climate Change-Cameroon (MECCOD), Ndop; Earth Journalism Network (EJN), Washington DC; Pan-African Parliamentarians Network on Climate Change (PAPNCC); Climate Change Radio Network (CCRN)-Bamenda as well as the Pan-African Climate Justice Alliance (PACJA) amongst others.

The media houses in the Northwest Region of Cameroon have interactive programs on climate change and other environmental issues. During such live broadcasts, the public is engaged through phone-in and text messages to contribute to the issues at stake as well as to ask questions and get clarifications from the resource persons in the studios. This is an indication that the public is very receptive to climate change messages. The radio stations broadcast a variety of climate and environmental related programs (Table 6.2). They have more than one climate and environmental related programs. The most popular of the programs is *'You and the Environment'.* When asked to the media organs why 'You and the Environment', they explained that the slogan is easy and straightforward to understand by the public.

Table 6.2: Climate-Environment related Programs on the Media in Northwest Cameroon

Media House	Climate-Environment programs
CRTV Northwest, Bamenda	Nature
Santa Meteo Radio	Climate & Environmental Slots
Ndefcam Radio, Bamenda	Environmental Conservation
Cameroon Baptist Radio, Bamenda	You & the Environment
Rush FM Radio, Bamenda	Daily Weather Forecast
Belo Community Radio	Water Forum, My Community & I, Famine Corner
Bui Community Radio, Kumbo	Environmental Protection, Bui Development Forum
The GrassLander Newspaper, Kumbo	Environment & Agriculture, Students' Corner
Oku Rural Radio	Focus on our Environment for our Tomorrow, Conserve Kilum Forest
City Community Radio, Kumbo	You & the Environment
Afoni Community Radio, Tatum	Environmental Issues
Savannah Frontier Radio, Ndu	Climate Change, You & Your Environment, Agriculture-Our Calling
Renaissance Community Radio, Ndu	Emergence, Apiculture, Gardening, Organic Manure
Voice of Moghamo, Batibo	You & the Environment, Water is Life
Savanna Frontier Radio Network, Nkambe	Climate Change, Farmers' Corner
Donga-Mantung Community Radio, Nkambe	Our Environment, Farmers' Show
Radio Evangelium, Kumbo	Reflections on Pope Francis' *Laudato Si'*, Agricultural Training Program (ATP), Understanding Integral Development
Jakiri Community Radio	Environmental Education, Farmers' Corner
Stone FM Community Radio, Ndop	You & Your Environment, Farmers' Corner, Hygiene & Sanitation, Women & Changing World
Abakwa FM, Bamenda	The Green Environment
Radio Hot Cocoa, Bamenda	You & the Environment
Radio Evangelium, Bamenda	Reflections on Pope Francis' *Laudato Si'*
The Eye	Climate Messenger
Eden Newspaper	Green Corner
Boyo Community Radio, Fundong	Trees for Life
Foundation Radio, Bamenda	–
Rainbow Radio, Mbengwi	–
Sky Sports Community Radio, Bamenda	–
Sky FM, Ndu	–

Source: Fieldwork, May-August 2016

'You and the Environment' is aired by Stone FM Community Radio Ndop (Ngokentugia Division), Voice of Moghamo Batibo (Momo Division), Savannah Frontier Radio Ndu (Donga-Mantung Division), Radio Hot Cocoa Bamenda (Mezam Division), City Community Radio Kumbo (Bui Division) and Cameroon Baptist Radio Bamenda (Mezam Division). In addition, three radio stations-Cameroon Baptist Convention Radio Bamenda, City Community Radio Kumbo, and Afoni Community Radio Tatum indicated that this program handles a variety of issues including climate change. Two stations-Savanna Frontier Network Nkambe and Savannah Frontier Radio Ndu, both in Donga-Mantung Division transmit programs entitled *'Climate Change'*. Other programs include *'Environmental Issues'* by Afoni Community Radio Tatum (Bui Division); *'Emergence'* on the airwaves of Renaissance Community Radio Ndu (Donga-Mantung Division); *'Our Environment'* via Donga-Mantung Community Radio Nkambe (Donga-Mantung Division); *'Environmental Education'* by Jakiri Community Radio (Bui Division); *'Farmers' Corner'* by Stone FM community Radio Ndop, Donga-Mantung Community Radio, Savanna Frontier Radio Network Nkambe, Savannah Frontier Radio Ndu and Renaissance Community Radio Ndu. *'The Green Environment'* is broadcast by Abakwa FM Bamenda (Mezam Division). *Reflections on Pope Francis' Laudato Si'* is a broadcast of the two Catholic radio stations-Radio Evangelium Bamenda (Mezam Division) and Radio Evangelium Kumbo (Bui Division). In addition, Radio Evangelium Kumbo broadcasts an all-encompassing program called *Understanding Integral Development.* The program addresses climate change, water resources management and land degradation amongst other unsustainable environmental practices. Four stations air programs on environmental conservation. These programs are *'Environmental Conservation', 'Conserve Kilum Forest' 'Environmental Protection', 'Trees for Life'* by Ndefcam Radio Bamenda (Mezam Division), Oku Rural Radio, Bui Community Radio (Bui Division) and Boyo Community Radio (Boyo Division) respectively.

The regional radio station of the Cameroon Radio and Television (CRTV) Northwest runs a weekly program- *'Nature'*. This program handles a cross-section of environmental issues such as climate change, water resources, environmental impact assessment and urban waste disposal in the city of Bamenda amongst others. This is a weekly live-studio broadcast with resource persons to handle the themes for each program. The presenter of the program in collaboration with the Regional Delegations of Agriculture and Rural Development, Environment, Nature Protection and Sustainable Development carry out field investigations across the entire region on climate change vulnerability, mitigation, and adaptation. CRTV Northwest also prepares and broadcast special packages for the following international days because they relate to climate and the environment: world water day, world meteorology day, earth day, international day for biological diversity, world environment day, world day to combat desertification and droughts, international day for the preservation of the ozone layer, world tourism day, world habitat day, international day for natural disaster reduction, world food day, world cities day, international day for prevention of the exploitation of the environment in armed conflicts and international mountains day. Although not all radio stations run programs on climate and the environment, they still have news items on these themes, especially during international days like world water day, earth day, world food day and others.

Three newspapers-The GrassLander in Kumbo (Bui Division), The Eye (Mezam Division) and Eden Newspaper (Regional Bureau, Bamenda, Mezam Division) that participated in this investigation run columns-*'Environment and Agriculture'*; *'Climate Messenger'* and *'Green Corner'* respectively. The GrassLander also runs a *Students' Corner* where secondary students report activities of their schools' environmental clubs.

The media houses also assign their staff to carry out field investigations of some physical evidence of the changing climate. Some of these field investigations include degraded watersheds, drying water bodies, recurrent droughts in agricultural lands, especially at *hollow frontiers* (fertile agricultural lands where farmers move seasonally), areas of unsustainable agricultural practices as well as other anthropogenic-driven environmental changes. In order to effectively pass across the messages, use is made of local languages and Pidgin English to communicate the effects of climate change based on local realities. The outcome of these field investigations culminates in the production of radio documentaries on local vulnerabilities and indigenous adaptations to climate change.

The Network of Environmental Stakeholders Cameroon (NEST-Cam) has formed an alliance of Non-Governmental Organisations and Common Initiative Groups to join efforts in fighting climate change (Table 6.3).

Table 6.3: Network of Environmental Stakeholders Cameroon (NEST-Cam)

NGOs/CIGs/Associations	Activities
The Greens	Environmental Conservation
Community Agriculture & Environmental Protection Project (CAEPP) Cameroon	Sustainable Agriculture & Environmental Protection
Save Your Future Association (SYFA)	Sustainable Agriculture
Association for Environmental Education (ASEC)	Environmental Education in schools
Sustainable Run for Development (SURUDEV)	Empowering the society to fight climate change
Centre for Indigenous Resources & Development (CIRMAD)	Natural Resources Conservation & Management
Rural Women Centre for Education & Development (RuWCED)	Women empowerment & Environmental Protection
Centre for Environmental Protection & Community Support Initiatives (CEPCSI)	Environmental Protection, Community Based Climate Change Adaptation
Association for Sustainable Development & Environmental Protection (ASDEP)	Environmental Protection
Society for Initiatives in Rural Development & Environmental Protection (SIRD-EP)	Rural Development & Environmental Protection
Mixed Farming Common Initiative Group (MIFACIG)	Mixed Farming
Kedjom-Keku	Protection of endangered Mountain Ecosystems of the Bamenda Highlands
Better World Cameroon	Permaculture
Community Initiative for Sustainable Development (COMINSUD)	Women & Climate Change, Environmental Protection, Peace Building, Rural Development
Y'Ems Group	Web company that donates 10% of profits to fight climate change
Alliance Française Cameroun, Bamenda	Provides venue for NEST-Cam activities
La Liberté Arts Group	Dramatizing environmental issues

Source: Fieldwork, May–August 2016

Common themes that reflect in the visions and missions of these associations are climate change mitigation and adaptation, climate change vulnerability assessment, environmental sustainability, livelihoods improvement, sustainable agriculture, enhancement of ecosystem services and many more. NEST-Cam works in collaboration with the Regional Delegation of the Environment, Nature Protection and Sustainable Development for the Northwest Region of Cameroon, the Bamenda City Council and other municipal councils within the city of Bamenda. NEST-Cam intends to spread her activities to other divisions of the Northwest Region and the entire country as a whole.

Some of the climate and environmental programs in radio stations are run by NGOs. For instance, *'The Green Environment'* on Abakwa FM is produced and aired by *The Greens.* The program *Nature* on the regional station of the CRTV is anchored in collaboration with some NEST-Cam members. *'Trees for Life',* broadcast over Boyo Community Radio Fundong in produced and presented in collaboration with *CIRMAD* and *SURUDEV.* The presenters of this program are *CIRMAD* and *SURUDEV* staff. Again, the program *'You and the Environment'* over Radio Hot Cocoa is produced and presented by *SURUDEV.*

Other NGOs actively involved in climate change and environmental communication are *Green Care Association,* Shisong *(GCA), Strategic Humanitarian Services (SHUMAS)* and *Cameroon Gender and Environmental Watch (CAMGEW).* The mission of *GCA* mission is to empower local communities towards sustainable use of local resources and to build networking opportunities across the world, with an overriding commitment to education, poverty alleviation, and community development. *GCA* promotes environmental sustainability through sustainable agriculture, watershed protection through analogue forestry, non-conventional livestock rearing, ecotourism, mushroom production, tree nursery amongst others. *SHUMAS,* on the other hand, promotes environmental protection and management through eucalyptus replacement, watershed protection and bio-farming, all in a bid to mitigate the effects of climate change in the Bamenda Highlands of Cameroon. *SHUMAS* has replaced 2 million eucalyptus trees with 2.5 million environmentally friendly trees. They have also planted watershed trees and life fencing to protect more than 70 water catchments. *CAMGEW* promotes the conservation and management of endangered mountain ecosystems.

Additional stakeholders in climate change and environmental communication in the Northwest Region of Cameroon include the Regional Delegation of the Environmental, Sustainable Development and Nature Protection for the Northwest, Regional Delegation of Forestry and Wildlife for the Northwest, Justice and Peace Commission of the Catholic Diocese of Kumbo, Regional Delegation of Agriculture and Rural Development for the Northwest and municipal councils.

It is also worth noting that *Stone FM Community Radio Ndop* was created by the *Media Synergy for the Promotion of Biodiversity Conservation and Community Development against Climate Change-Cameroon (MECCOD).* This radio station produces a variety of climate-environment related programs such as *'You and Your Environment', 'Farmers' Corner', 'Hygiene and Sanitation', 'Women and Changing World'*, amongst others. This goes a long way to buttress the synergy between the media, Non-Governmental Organisations and Common Initiative Groups in promoting climate change and environmental communication. There is also a remarkable collaboration between media houses in producing and broadcasting climate-environmental programs. For instance, some editions of the program *'Emergence'* are co-productions and presentations of *Renaissance Community Radio Ndu* and *Savannah Frontier Radio Ndu.*

Limitations to Climate Change Communication

Climate change is a challenging issue to convey because the long-term average changes set in slowly and imperceptible (Moser and Dilling, 2011). Changes in medium-term and long-run have emerged only recently from the daily, seasonal and inter-annual variations. In the same way, not all media houses in the Northwest Region of Cameroon are taking climate and environmental issues seriously, especially fewer climate disaster narratives (Banda *et al.*, 2013). Some of these radio stations were hesitant to participate in the study with the argument that climate change is caused by industrial emissions in the developed countries. Others who participated in the survey do not run any programs on climate change because of inadequate resource persons to tackle the subject and their inability to hire specialists in climate and environmental issues. These include Sky Sports Community Radio Bamenda, Rainbow Radio Mbengwi, Sky FM Ndu and Foundation Radio Bamenda (Table 6.4). Again, the general reporting on climate and environmental issues is inadequate (Table 6). Apart from world water day, world environment day and world food day, the rest of the international days related to climate and the environment are under-reported. One reason for this is that the media outlets do not know the dates which these international days are commemorated.

Table 6.4: Reporting Climate and Environmental International Days

International Day	Date	Freq	%
World Water Day	March 22	29	100.0
World Meteorology Day	March 23	03	10.34
Earth Day	April 22	13	44.82
International Day for Biological Diversity	May 22	07	24.13
World Environment Day	June 5	22	75.85
World Oceans Day	June 8	01	03.44
World Day to Combat Desertification and Droughts	June 17	12	41.37
International Day for the preservation of the Ozone Layer	September 17	08	27.58
World Tourism Day	September 27	10	34.48
World Habitat Day	October 3	10	34.48
International Day for Natural Disaster Reduction	October 12	09	31.03
World Food Day	October 16	17	58.62
World Cities Day	October 31	04	13.79
International Day for Prevention of the exploitation of the Environment in Armed Conflicts	November 6	08	27.58
International Mountains Day	December 11	03	10.34

Source: Frequencies and percentages computed from field data (May-August 2016)

Some of the radio stations are not living up to expectation for the purpose for which they were created. For instance, Santa Meteo Radio (Mezam Division) does not broadcast daily weather forecast. They also do not know the date which world meteorology day is commemorated. However, they have news items on the risks of climate change from time to time. Santa Meteo Radio is only aware of world water day, international day for the preservation of the ozone layer and world food day.

Another shortcoming of the media houses is the mastery of the concept of climate change. A significant proportion (≈70%) of the staff in these media houses have limited knowledge of the science of climate and environmental change. Thus, there is a need for capacity building and training. Other factors, especially policy options and resistance to change from indigenous communities come into play. Thus, for communication to be effective in leading to active engagement, it must be supported by policy, economic and change of mindset of the people that will pave the way for the concerns and good intentions to be achieved. On the part of NGOs, they have lofty ideas on paper, web blogs and social media. When one gets to see how they implement these ideas on the field, there is very little. Many are yet to understand the effects of climate change at the local level. It is from there that they will be able to progress with their agendas of climate change mitigation and adaptation. Like the media, many NGOs in the Northwest Region of Cameroon do not commemorate all the environment-related international days because they do not know the dates. All these culminate in the fact that media outlets in the Northwest Region of Cameroon have little knowledge of environment-related international days (Fig. 6.2).

Fig. 6.2: Decreasing trend of media awareness of International Environment Days

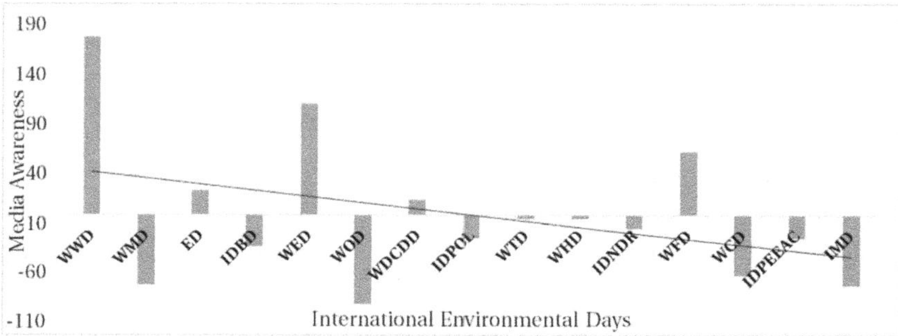

WWD	World Water Day
WMD	World Meteorology Day
ED	Earth Day
IDBD	International Day for Biological Diversity
WED	World Environment Day
WOD	World Oceans Day
WDCDD	World Day to Combat Desertification and Droughts
IDPOL	International Day for the preservation of the Ozone Layer
WTD	World Tourism Day
WHD	World Habitat Day
IDNDR	International Day for Natural Disaster Reduction
WFD	World Food Day
WCD	World Cities Day
IDPEEAC	International Day for Prevention of the exploitation of the Environment in Armed Conflicts
IMD	International Mountains Day

Source: Computed from percentages in Table 7

That is why the trend of media awareness of international environment days is decreasing. Thus, the coefficient of determination (R^2) is as low as 14.5%. The rest of the international climate-environment related days (World Meteorological Day, International Day for Biodiversity, World Oceans Day, International Day for Preservation of the Ozone Layer, International Day for

Natural Disaster Reduction, World Cities Day, International Day for Prevention of the exploitation of the Environment in Armed Conflicts and International Mountains Day remain relatively unknown to the media organs. This explains why the media in the Northwest Region of Cameroon produce and broadcast programs only on World Water Day, Earth Day, World Day to Combat Desertification and Droughts and World Food Day. The rest of the international days remain unknown to the media.

Conclusions and Future Prospects

Climate change communication is gaining ground in the Northwest Region of Cameroon. All the media houses surveyed were fully aware of the vulnerabilities and direct impacts of climate change around their local environment. Faced with this evidence, climate communicators have been engaging the public in the fight against climate change through the following strategies:

- They attempt to increase public understanding of the prevailing environmental conditions and provide more information on the assumption that inadequate knowledge is the major stumbling block to action.
- Fear strategy is used to provoke positive action, but often, this achieves the opposite effect.
- Climate change communication is based on the credibility of the authorities like the IPCC and other climate scientists. The local media insists and make it clear that their reporting is based on scientific evidence of climate change as the most convincing story, irrespective of the differences among audiences.
- The local media in the Northwest Region of Cameroon try to reach the masses through traditional communication channels. Communication on environmental and climate change is only part of the picture. Raising awareness and discussing issues does not directly result in behaviour change or policy action.

There is a need for more interactive radio programs on climate and the environment especially for the media houses that do not air such programs. Journalists need more capacity building and training on climate change and environmental reporting. Journalism clubs should be encouraged in schools where students can disseminate environmental issues to their peers. Radio quizzes should be aired during all climate and environment-related international days. Socio-cultural, religious gatherings like churches should be used as communication channels on the dangers of unsustainable activities like poor agricultural methods. During international days, sketches and poems should be used as a means to get across climate change messages. Hands-on-the ground projects like tree planting should be encouraged through communities and NGOs. Traditional address systems *(Town Cryer)* should be used to sensitize the rural populations on the dangers of their activities like overgrazing, slash and burn farming and bushfires. Newspapers have a duty to dedicate columns for climate and the environment. Use of local languages to reiterate causes of climate change, vulnerability and adaptation should frequently be aired through radio stations. Posters in public spaces on the dangers of climate should take precedence. Sensitisation and workshops at grassroots should be encouraged. The youth should be fully engaged in climate change communication through popular social media and community forums.

Acknowledgements

We wish to thank Dr. Louise-Marie Bégné, (Regional Delegate of Communication for the Northwest Region of Cameroon) and Mr. Gilbert Fonyuy, (Regional Chief of Public Communication for the Northwest Region of Cameroon) who issued us an ethical clearance to collect data from media houses.

References

African Development Fund (ADF), 2013, Grassfield, Rural Infrastructure and Participatory Development Support Project, Phase II. Grassfield Decentralized and Rural Development Project II (GP-DERUDEP II), Bamenda, 9.

Banda, F., Shanahan, M., Shubert, W., Scherer, C., & Corcoran, T., 2013, Climate Change in Africa: A Guidebook for Journalists. UNESCO, Paris, 12.

de Wit, S., 2011, Global Warning: An Ethnography of the encounter of Global and Local Climate Change, Discourses in the Bamenda Grassfields, Cameroon. African Studies Centre, Leiden, 139.

Department for Communities and Local Government, 2007, Climate Change Communication Strategy: A West Sussex Case Study. European Spatial Planning Adapting to Climate Events, 2.

Egan, A., 2013, Knowledge Management Strategy on Climate Change Adaptation for Cameroon. Ministry of the Environment, Protection of Nature and Sustainable Development, Yaoundé, Cameroon, 17-19.

Filho, W.L. 2009, Communicating Climate Change: Challenges ahead and action needed. International Journal of Climate Change Strategies and Management. 1, (1), 6-18.

Intergovernmental Panel on Climate Change (IPCC), 2016, Meeting Report of the Intergovernmental Panel on Climate Change Expert Meeting on Communication. Lynn, J., M., Araya, Ø., Christophersen, I., El Gizouli, S.J., Hassol, E.M., Konstantinidis, K.J., Mach, L.A., Meyer, K., Tanabe, M., Tignor, R., Tshikalanke, J-P., van Ypersele (Eds.), 2016, World Meteorological Organization, Geneva, 229.

Intergovernmental Panel on Climate Change (IPCC), 2014, Climate Change 2014: Impacts, Adaptation and Vulnerability, Contribution of Working Group II to the IPCC Fifth Assessment Report. Cambridge University Press, Cambridge, United Kingdom and New York, 1774.

Lambi, C.M., 2001, The Impact of Human Activity on Land Degradation in some Highland Regions of Cameroon: Implications for Development. Environmental Issues: Problems and Prospects. Unique Printers, Bamenda, 53.

Molua, E.L., & Lambi, C.M., 2007, The Economic Impact of Climate Change on Agriculture in Cameroon. Policy Research Working Paper 4364, World Bank, Washington, DC, 33.

Monroe, M.C., Bode, C.L.L. & Megalos, M.A., 2015, Challenges in Communicating Climate Change to Extension Audiences. University of Florida, 5.

Moser, S.C. 2010, Communicating climate change: history, challenges, process and future-e directions. John Wiley and Sons, New York, 23.

Moser, S.C., & Dilling, L., 2011, Communicating Climate Change: Closing the Science-Action Gap. In John S, Dryzek, Richard, B., Norggard, David Schlosberg, (Eds.), 2011, Oxford Book of Climate and Society. Oxford University Press, Oxford, 161-174.

Myhre, G. D., Shindell, F-M., Bréon, W., Collins, J., Fuglestvedt, J., Huang, D., Koch, J-F., Lamarque, D., Lee, B., Mendoza, T., Nakajima, A., Robock, G., Stephens, T., Takemura, & H., Zhang, 2013, Anthropogenic and Natural Radiative Forcing. Climate Change 2013: The Physical Science Basis. Contribution of Working Group I to the Fifth Assessment Report of the Intergovernmental Panel on Climate Change [Stocker, T.F., D., Qin, G-K., Plattner, M., Tignor, S.K., Allen, J., Boschung, A., Nauels, Y., Xia, V., Bex, P.M., Midgley, eds., 2013]. Cambridge University Press, Cambridge, United Kingdom and New York, NY, USA, 675, 686-687.

Nerlich, B., Koteyko, N. & Brown, B., 2009, Theory and Language of Climate Change Communication. John Wiley and Sons, New York, 17.

Ngonga, L.N., & Tume, S.J.P., 2015, Indicators of Spontaneous Settlements in Bamenda, Northwest Region of Cameroon. African Journal of Social Sciences, Unique Printers, Bamenda, 6(4) 97.

Northwest Region (Cameroon), https://en.wikipedia.org/wiki/Northwest Region_(Cameroon). Accessed on June 25, 2016.

Shome, D. & Marx, S., 2009, The Psychology of Climate Change Communication: A Guide for Scientists, Journalists, Educators, Political Aides and the Interested Public. Centre for Research on Environmental Decisions, New York. 45.

United Nations Development Program (UNDP) 2009, Supporting Integrated and Comprehensive Approaches to Climate Change Adaptation in Africa–Cameroon. Ministry of the Environment, Protection of Nature and Sustainable Development, Yaounde, 101.

United Nations Framework Convention on Climate Change (UNFCCC), 2006, UNFCCC Handbook. UNFCCC Secretariat, Bonn.

Ward, B. & Menezes, S., 2008, Communicating on Climate Change: An Essential Resource for Journalists, Scientists and Educators. Metcalf Institute for Marine & Environmental Reporting, Narragansett, 27.

Chapter Seven

TARGETING AUDIENCES AS IMPORTANT GAME CHANGERS IN THE CLIMATE CHANGE COMMUNICATION PUSH: EXAMPLES AND CASES

Innocent Chirisa, *University of Zimbabwe, Zimbabwe*

Abraham R Matamanda, *University of the Free State, South Africa*

John Mutambwa, *University of Zimbabwe, Zimbabwe*

Abstract

Too often, good initiatives or important subjects fail to take root because the message and crux of the matter are little understood by the audience they are intended for. This explains why very important policies, plans, programmes, and processes never get to the desired efficacy levels. Disappointingly, good intentions 'die' without making the necessary impact because the majority of the targeted audiences, particularly individuals, households, and communities never get to receive the messages or rather wrong platforms and media are often used. At the present moment, the message of climate change is central to the development discourse everywhere. However, the reception of the message varies from community to community, given the mix of the audiences which include sceptics, heretics and the little-informed who form the biggest group especially in Africa. The aim of this chapter is to identify, explain and discuss dimensions in and methodologies towards 'reaching out' to individuals, households and communities in a bid to provide an informed space especially in preparing communities to adjust to the vagaries of climate change which include flooding, droughts, heat waves and the general water stress in both urban and rural settings. The study uses examples and cases from Zimbabwe to concretise issues in the debate.

Key words: Audience • Climate Change • Mass Communication • Zimbabwe

Introduction

Climate change is a reality that is proving to be a key policy agenda at global level (IPCC 2013). Various efforts have been made by the global world to address climate change issues in all regions across the world (UNESCO, 2014). Climate change is often perceived as a scientific and technical problem that requires technical language and expertise to explain and understand (Hiwasaki *et al.*, 2014). Therefore, the dissemination of climate change information in most instances is not conveyed in a manner that is acceptable and readily understood; a manner which can significantly change behaviour or prompt adaptation and mitigation strategies among audiences, especially those in the developing world and rural communities in particular. The assumption, then, is that climate change communication is complicated by the fact that accurate communication of the 'knowns' as well as the 'unknowns' might require rather *technical language* which can be interpreted by a few individuals and stakeholders, hence the information that is conveyed fails to serve its intended purpose as it tends to be too complex to decode for the large segments of society (European Environmental Agency-EEA, 2016). The target audience, on the other hand, is constituted of heterogeneous groups of people who have diverse interests, needs, values, and norms. This results in the policies, plans, and strategies aimed at climate change adaptation and mitigation failing to achieve their intended goals of effecting change or being accepted by specific target audiences. Moreover, good intentions die without reaching the intended audiences possibly due to the use of inappropriate language or inappropriate media (UNESCO, 2014). Thus, it is always important to formulate communication tools that will enable the dissemination of climate change information to specific audiences so as to facilitate the transmission of the information as well as promote its acceptance by the target audience (Monroe *et al.*, 2000). It is in this vein that this study seeks to explain and discuss dimensions in, and methodologies for 'reaching out' to individuals, households, and communities in a bid to provide an informed space, especially in preparing communities to adjust to the vagaries of climate change. The effects of climate change include flooding, droughts, heat waves and the general water stress in both urban and rural settings.

In order to achieve the broad aim of the study, the objectives of the study are to:

- Identify the media and strategies that are used to disseminate climate change information in Zimbabwe as well as in targeting the audience;
- Analyse the nature of the information that is conveyed to particular audiences;

- Examine the perceptions of the audience towards the information they receive;
- Evaluate the effectiveness of the information received by certain target audiences in contributing to climate change adaptation and mitigation.

The structure of the chapter is set as follows: Literature Review and Theoretical Framing section which focuses on the review of relevant literature pertaining to climate change communication, perceptions, and expectations of the audiences and models that may be used for communication. The research methodology section explains and justifies the research design for the study, data collection methods, and the data analysis approaches. The findings of the study are then presented in the Results section which is then followed by the Discussion and Policy sections. Finally, the conclusion section winds up the study.

Literature Review and Theoretical Framing

The focus of this section is on the review of related literature and theoretical framing that guide this study. The section begins by reviewing the literature on climate change and climate change communication and then introduces the issue of mass communication and the theories that inform this paper. The section goes on to review literature that pertains to the barriers to climate change communication. The tools for effective climate change communication are then presented at the end of the section.

Climate Change

Climate change has different meanings to different sectors of society (Otto-Banaszak *et al.*, 2010). Of utmost importance is the fact that climate change is not easily understood by the layperson in society, mainly because it is a gradual process whose results are not instantly visible (Weber, 2010). On the other hand, insufficient conceptualisation and trust from the audience towards climate change information also complicate the transfer of scientific descriptions of climate change information from scientists to the intended audiences. From studies on climate change in most regions across the world, it emerges that climate change communication is also affected by norms, values, cultures, and religions of the local communities which, in most instances, determine the level of acceptance of any adaptation and mitigation strategies and policies designed for the community (UNESCO, 2014). In other instances, there is simply lack of information on climate change, following failure by communities to document proceedings from meetings on climate change (Mudombi, 2013). The result is that the concept of climate change remains elusive and ill-understood by the layman, yet this information is so vital for the success of adaptation and mitigation strategies to climate change.

Mass Communication

Dissemination of climate change information is always intended for large audiences; hence climate change communication comes under the rubric of mass communication. Okenwa (2002) defines mass communication as the transmission of information and ideas to a largely dispersed, heterogeneous audience simultaneously. It is interesting to note that a target audience is never a passive group who will simply accept whatever information is conveyed to them. It is in this vein that Flower (2008) explains the diverse opinions held by a particular audience which results in the acceptance or rejection of certain information. The Dissonate Theory of mass communication thus helps to explain why mass media does not always have persuasive or conversion effects on their audience (Okunna, 1999). The target audience may simply decide to reject the information that will be conveyed to them, hence the need to be very cautious when it comes to transmission of some information. This situation also applies in the realm of climate change where there are many myths and beliefs that appear to complicate the transfer of technical knowledge on climate change that has to be conveyed to a particular audience.

Various problems have been identified in climate change communication. According to UNESCO (2014), the use of technical terms and acronyms by climate change agencies greatly hinder the effectiveness of information as the audiences often fail to decode the information. This follows findings from a study in the Pacific where most media complained that they found it difficult to convert press releases on climate change into compelling stories due to the technicality of the messages (UNESCO, 2014). Furthermore, press releases were not localised and did not narrate personal accounts and stories about climate change. The result is that the target audience felt that the information being conveyed to them was alien and thus failed to have an impact on them (Chagutah, 2010).

Moser (2010) identifies a set of factors that mar the success of climate change communication. These factors are discussed hereunder. First, there is the fact that the causes of climate change are invisible, hence it is difficult for the audience to take heed of communication on climate change that encourages them to adopt mitigation and adaptation strategies. Second, the impact of climate change is often distant. In this regard, the impact of climate change is both spatially and temporally distance; which means that the impact may not be observed at that particular time or it may be too small to notice. Thus, such a situation breeds a lackadaisical attitude towards climate change which will be viewed as just a myth. Third, people in the contemporary world are usually well insulated from the environment to such an extent that they do not recognise or may fail to notice the changes in climate. We are living in a highly technological world where buildings are well-ventilated, the environment is heavily altered (for example, cloud seeding to produce rain) and much is constantly done to regulate the environment. All this obscures climate change experience and reality to the ordinary person. Fourth, climate change is very complex and uncertain. Moser (2010) maintains that 'most individuals (even scientists) cannot and will never fully grasp and hold this amount of scientific complexity and uncertainty in their minds, much less be able to process it systematically. And even if individuals accepted the overarching scientific consensus statements as valid, there is additional complexity and uncertainty about the technically feasible, environmentally benign, economically affordable, and morally preferable mitigation (and adaptation) responses'. Fifth, in some instances and areas, there are inadequate signals indicating the change in climate processes and variability. Finally, the other dimension looks at self-interest, justice and humanity's common fate. Very often, humans have acted in selfish ways, and this also applies to the realm of climate change communication where information may not be conveyed simply because some officials feel their businesses or activities may be at stake. Or rather, it may be an attitude of Not in My Back Yard (NIMBY).

If climate change communication is to become successful through increasing awareness, informing, or changing behaviour in the target audiences, it has to be designed, based on the forthcoming model by Monroe *et al.* (2000, p.5). The model considers four aspects in relation to environmental communication in which climate change communication also fits in. These aspects are Goal, Audience, Medium, and Message, which must be looked at in the order presented here.

- ***Goal***: relates to the need to set a clear goal(s) which spell(s) out what the target audience wants to do, what type of information has to be conveyed and what aspects of their lives have to be changed by the information.
- ***Audience***: Moser (2010) argues that attention paid to the audience needs and differences in climate change communication has been limited historically. Thus, the target audience has to be selected, and consideration has to be given as to how the environmental issue at hand impacts on their lives. The utility of considering the needs of the audience is clearly espoused by Rose *et al.*, (2005) who opine that effective climate change communication ought to tap into the motivations of the target audiences.

- ***Medium***: There is then need to identify the audience's 'media diet' so as to ascertain where the target audience gets its information from. The medium presents the platform through which information will be conveyed to the target audiences. In other words, the media act as the messenger. Selecting the most befitting medium is key in climate change communication because just as society is heterogeneous, so are the preferences for communication media. Options for communication media may include TV, radio, newspapers, social media, church, academics, traditional leaders, and politicians.
- ***Message***: Finally, the message has to be framed after taking into consideration the goal, audience, and medium so as to reach and be accepted by the target audience. This is also one other critical part of climate change communication which must be approached with caution. The central question revolves around the nature and type of message that has to be communicated to the people. The most critical part of a message is that it must have the ability to keep the attention of the target population, lest it will be ignored. O'Neill and Nicholson-Cole (2009) maintain that the tone of a message is of utmost importance as it can help in conveying the message in a manner that can, for example, induce fear in the audience.

Tools for Better Communication by EEA (2016)

Nudging is a strategy that helps to influence the behaviour of the target audience in such a way that they may change their behaviour as a result of certain information. In essence, nudges are not mandates but rather incentives, motivation or assistance that will persuade the target audience to change their behaviour when they see a particular message (Thaler and Sunstein, 2008). Lindahl and Stikvoot (2015) argue that interventions for nudging ought to be easy and cheap and must help by all means possible to maintain freedom of choice. That being the case, nudging will help to disseminate information that is lucrative to the audience such that change in behaviour becomes imminent. This will eventually help in efforts aimed at climate change adaptation and mitigation. Examples of nudges that have been used in some policy areas include the use of tools such as bans, fines and traditional information campaigns (EEA, 2016).

Social marketing is, by definition, a process that makes use of marketing principles, tools and techniques to create, communicate and deliver value in order to influence the behaviour of target audiences in relation to issues that will benefit the audience (Kotler and Lee, 2008). In social marketing, the main objective again is to change the behaviour of a particular audience towards something which will benefit the audience. According to EEA (2016), audience segmentation may be used where a particular audience is targeted, and those responsible for conveying the information will develop tailor-made messages, information packages and incentives that will eventually conform to the expectations of the target audience.

Public participation refers to the engagement of those individuals who are affected by a particular issue in the decision-making process. Such a process is often associated with input from the public who do not simply emerge as recipients of decisions. In communication, public participation also has a place. Most communication media are one way and allow information to be disseminated to the target audience with little or no feedback from the audience. Examples of such media include newspapers, radio, and television. For climate change communication to be effective, the communication process has to be a two-way process. The audience must also have a voice in the communication process when it comes to climate change communication. In the context of communication, public participation comes in the rubric of citizen journalism. Lasica (2003) identifies a number of ways which constitute citizen journalism, and these include audience participation, photos or videos recorded by mobile phones, letters from readers and private video broadcasting sites. The utility of citizen journalism is that people end up being both consumers and producers of news and they eventually give credit to the news and information that will be produced (Gillmor, 2004).

Social media refers to communication that is interactive in nature and is web-based as well as supported by mobile technologies. To some extent, social media may be categorised as citizen journalism because the audience gets to create and debate on certain issues. In this millennia social media has come to be a force to reckon with, regarding the transmission of information to various people. Information is shared, discussed and feedback is given, which helps the sender to get feedback from multiple recipients. As a result, social media is one other important communication tool that may help to increase the visibility and acceptance of climate change information by the target audience (Baruah, 2012).

Methodology

The study is qualitative in nature and uses the phenomenological approach. The phenomenological approach seeks to illuminate the specific, identify phenomena through how they are perceived by the actors in a situation (Lester, 1999). The phenomenological approach refers to the way participants experience things and what these experiences mean to the particular participants.

The phenomenological approach has been used for this study because it helps the authors to understand the social phenomenon from the audience's own perspectives wherein they get to describe the how they relate to climate change information that is communicated to them. Data was thus collected using key informant interviews that included officials from Ministry of Environment, Water and Climate (MEWC), EMA, school children and residents in both rural and urban areas. An intense analysis was also done on documents such as Climate Change Policy, Environmental Management Act and Water Policy. The official websites of EMA, MEWC and other NGOs engaged in climate change information were also visited. Once the data was collected, it was analysed using content and text analysis where it was put into themes that were then analysed by the authors for meaning.

Results: Cases from Zimbabwe

Zimbabwe is experiencing climate change as evident from the increase in climate extremes across the country. Of note are the increasing average daily temperatures and reduced rainfall. Predictions are that daily average temperatures will increase between 0, 5°C to 2°C by 2030 and 1°C to 3, 5°C by 2050 while some models predict a 10–20% decline in rainfall by 2050 (UNDP, 2011). What makes Zimbabwe more vulnerable to climate change is that over 60% of the country's population lives in rural areas and depend on rain-fed agriculture (ZimStat, 2012).

Various platforms and media are used to convey climate change information to different audiences in Zimbabwe. Considering that the majority of the country's population is concentrated in rural areas and engage in agricultural activities, local farmers become the most prominent audience in the rural areas. Rural communities are more vulnerable to climate change than their urban counterparts hence much of the communication is targeted at them. However, climate change is also prevalent in urban areas, which also calls for the dissemination of climate change information to urban communities in Zimbabwe. Chagutah (2010) observes that even Parliamentarians in Zimbabwe are not fully aware of climate change; hence these also emerge as a target audience. Learners in primary and secondary schools and tertiary institutions were also identified by an official from MEWC as a target audience to climate change information. What it means then, is that information has to be packaged specifically for each of these categories of the audience because the media and message can never be the same for all audiences if effective communication on climate change is to be achieved.

Among the Non-Governmental Organisations (NGOs) that are concerned with climate change issues in Zimbabwe is Regional Environment Organisation (ZERO), a regional NGO, based in Harare. Among its mandates is the task of providing updated information on climate change in Zimbabwe. ZERO uses a number of strategies to disseminate climate change information to various groups in Zimbabwe and these range from research articles, case studies, newsletters, annual reports and the internet. The information provided by ZERO is channelled to various audiences that include universities and research institutions, NGOs concerned with climate change issues, communities and selected housing media. The social media (Internet, WhatsApp, Twitter and Facebook) are a common platform used by ZERO to convey their messages in simple language as well as through nudging (Boxes 7.1 and 7.2).

Box 7.1: Climate Change Messages on the ZERO webpage

Climate Change: Presence of extreme weather events. Affects all sectors. Either we adapt or mitigate climate change.
Solar Power: Is clean, renewable and reliable. It reduces Green House Gas emissions. It helps to reduce energy reliance on fossil fuels.
Solar Energy: Fastest growing energy source. Free and abundant in Zimbabwe. Sustainable source of energy.
Wind Power: Is the fastest growing clean energy source in the world. It relies on the renewable power of the wind. It is a sustainable source of energy.

Source: (http://www.zeroregional.com/)

Box 7.2: Understanding local weather patterns

I. Make an effort to understand your local weather patterns and the nature of hazards that can affect your area.
II. Prevent drowning, supervise children at play near water bodies. In case of emergency, dial 112.
III. Never attempt to cross flooded rivers whether walking or driving. In case of emergency dial 112.
IV. You are in danger of floods if your home is close to a river, on a low lying area, or on a wetland, timely move to higher ground. In case of emergency dial 112.

The *National Climate Change Response Strategy* for Zimbabwe acknowledge that Zimbabwe is constrained by its limited ability to put in place, appropriate measures in order to respond to climate change because of lack of human, institutional and financial resources (Government of Zimbabwe, 2015). Thus, the strategy ledges to promote public awareness-raising and communication on climate change in the country through: implementation of a communication strategy for raising awareness on climate change; encouraging sharing of information and networking on climate change issues at the local, regional and national levels; and communicating climate change messages that will incorporate indigenous knowledge systems (Government of Zimbabwe,2015). It emerges that there is a need for capacitating communication channels that relate to climate change through raising awareness of climate change. Therefore, the first task, even before the message is conveyed to the audience, is to conscientise the target audience so that whatever information they receive will be significant to them. In this vein, the strategy goes on to propose the use of communication strategies such as posters and pamphlets but hasten to caution that the messages on such media should be written in simple language that can be easily understood by the audience as well as giving them the ability to relate to such information. This also confirms observations by Chagutah (2010) and Mudombi (2013) who maintain that, just like in most instances across the world, information on climate change in Zimbabwe is not available in formats and language appropriate to the respective target audiences such as policymakers, industry, commerce, extension officers, NGOs and the community at large. In the recently-launched climate change policy for Zimbabwe, the document is too technical and difficult for the layperson to interpret. Besides the *Climate Change Response Strategy* for Zimbabwe, there is a dearth of material to guide the development and implementation of climate change adaptation and mitigation strategies in the country, as well as for general climate change awareness.

The Ministry of Environment, Water and Climate looks into climate change issues in Zimbabwe and, as such, is the main institution responsible for the dissemination of climate change information to all stakeholders in the country. The ministry has a Climate Change Management Department that is responsible for climate change issues. The mission of the Climate Change Management Department in the MEWC is to climate-proof all socio-economic sectors of Zimbabwe through effective climate change management. This department is responsible for climate change information dissemination through publications and research. However, the department's website shows scanty resources that may significantly help in informing climate change among specific audiences. Considering the utility of social media in transmitting information, the department ought to capitalise on such a platform and possibly make it interactive so that those who have access to the internet can benefit from climate change information. Moreover, the resources appearing on the website seem to be too academic and scholarly such that they may be useful only to particular audiences, thus, side-lining out-of-school audiences such as the youth and illiterate who do not have access to such reading materials. This confirms observations by UNDP (2011) that there is limited technical expertise in climate change information packaging and systematic data collection in the country.

To reach this target audience (out-of-school youths and some of the illiterate yet most vulnerable groups), there is a need for the systematic collection and packaging of information that will be channelled to particular audiences. In this regard, the *National Climate Change Response Strategy* proposes the use of opinion leaders to disseminate information. Opinion leaders such as traditional leaders (chiefs, headmen, etc.), church leaders and politicians are instrumental figures who have to make use of their influence to influence behaviour and decisions on climate change adaptation and mitigation. Information may be conveyed during gatherings such as rallies, community meetings, and church services. Drama may be used on both radio and TV to inform a wider audience while in some instances road shows may also be useful in transmitting climate change information. However, the response strategy for Zimbabwe says little about making use of social media and ICT in climate change communication that has the potential to reach a wider audience, especially the younger generation who are active on social media such as WhatsApp, Facebook, LinkedIn and Twitter among many others.

Public participation in climate change communication in Zimbabwe seems to be very minimal. There are limited platforms where citizens engage in climate change communication. At times various stakeholders are invited to public meetings where information will be disseminated on climate change issues confronting the country. Examples are the launch of *Zimbabwe's Climate Change Response Strategy* where stakeholders were invited from different parts of the country. However, proceedings at such events are largely dominated by academics who fail to simplify the technical issues relating to climate change in spite of a diverse and heterogeneous audience. The result has been murmuring among the audience all the way as the seminar proceeded, which shows lack of concentration from the audience. Apparently, the message that was being conveyed did not relate much to them. In this case, it emerges that although efforts have been made to engage members of the public, the platform does little to support such initiatives as the members of the public end up simply as spectators with little contribution to the complex issues that are discussed.

Climate change communication in Zimbabwe is also stifled by poor coverage and capacity to report on climate change issues by the media. Chagutah (2010) observes that the two leading newspapers in Zimbabwe (*The Sunday Mail* and *The Standard*) do not fully address climate change issues at the local level; rather, what is contained in the newspapers are events at the global level and some extreme disasters such as heat waves and floods (Box 7.3).

Box 7.3: Climate Change Communication in Chiredzi

Chiredzi is one area that is worst affected by climate change in Zimbabwe. Climate change communication has been facilitated by different stakeholders who target different audiences in the region especially women. The main thrust of the messages has been on increasing resilience of communities to the effects of climate change. This has been through workshops and various other projects that aim to increase the level of awareness among the communities as well as the use of climatic information. Workshops have always been a successful way of disseminating information to these groups in that they are more interactive and the audiences get to be well-informed and educated in the end. An example is the project by UNDP–Coping with Drought and Climate Change in Zimbabwe.

In this regard, even the literate audiences who can read, especially those in urban areas who do not get to receive adequate information on climate change which would help to inform adaptation and mitigation strategies. Furthermore, the target audience also tends to perceive climate change as something that will not have a significant impact on Zimbabwe. In random interviews held with some citizens in Harare, it emerges that there are limited platforms form citizen journalism in Harare and Zimbabwe at large. Considering that the media fails to cover and report on climate change effectively, the same also applies to the layperson.

Discussions and Policy Options

There is a need for decentralising and synergising climate change resources and data in all areas across Zimbabwe. Resources and information pertaining to climate change must be made readily accessible to all communities in both urban and rural areas. In order to overcome the deficiencies in climate change information dissemination to the audiences and improve climate change communications, government (at all levels) and non-state actors need to develop a robust organisational and institutional mandate that will help to disseminate climate change information to the intended audiences. Such an arrangement will also help to overcome overlaps in information sharing from the various climate change agencies to the recipients (audiences). In as much as resources will be decentralised, it is of utmost importance to train and educate the public as well as media with regards to climate change issues. The media needs training so as to equip them with the relevant skills and expertise that will help them to develop and eventually disseminate interesting and locally-relevant stories pertaining to climate change. Furthermore, information will not be replicated, or rather some information will not be sidelined as they will be a well-coordinated and fully functional institutional set-up geared for climate change communication. Although climate change is scientific in nature, there is also great need to understand better social and cultural aspects of communication and technology relating to climate change communication. While designing the tools, methods and policies for climate change adaptation and eventually communication, there is need for the relevant authorities to consider community perceptions, culture, values, norms and environmental conditions as these have been found to have profound effect on the manner in which society gets to accept or reject whatever information is passed on to them.

Conclusion

Climate change communication targets specific audiences who, in most cases, are diverse and heterogeneous. That being the case, there is a need for communication on climate change to be packaged in such a way that it is accepted by the target audience who will, in turn, eventually change their behaviour. That being the case, messages have to be packaged in such a way that the intended objectives are achieved, which will contribute to the mitigation and adaptation to climate change. The study concludes that there are various factors which hinder the success of climate change communication in Zimbabwe and these usually result in the audience either not having access to the information or when they do get access to the information, it tends to be too technical and beyond their comprehension. Overall, there is a need for information on climate change in Zimbabwe to be packaged systematically taking into consideration the needs of the target audience. Furthermore, the wide contemporary communication platforms such as social media are not effectively utilised for climate change communication which also contributes to the ineffectiveness of the climate change communication in the country. Instead of effectively using such platforms to disseminate important information, communication on social media is never taken seriously in African countries such as Zimbabwe, rather it is mostly used for entertainment and amusement.

References

Baruah, T.D., 2012, Effectiveness of Social Media as a Tool of Communication and its Potential for Technology enabled Connections: A micro-level Study. International Journal of Scientific and Research Publications 2(2), 1-10.

Chagutah, T., 2010, Climate Change Vulnerability and Adaptation Preparedness in Southern Africa: Zimbabwe Country Report. Cape Town: Heinrich Boll Stiftung Southern Africa.

European Environmental Agency (EEA), 2016, Communication, Environment and Behaviour: A Scoping Study on the Links between Public Communication, Environmental Policy Implementation and Behavioural Science. London: European Environmental Agency.

Flower, J., 2008, Mass Communication and Audience Response. [Online]. Available from www.flowers.wrytestuff.com, Accessed on 15 January 2017.

Gillmor, D. (2004). We the Media, Grassroots Journalism by the People for the People. Beijing: O'Reilly Media Inc.

Government of Zimbabwe, 2015, Zimbabwe's National Climate Change Response Strategy. Harare: Printflow.

Hiwasaki, L., Luna, E., Syamsidik, S., & Shaw, R. 2014, Local & indigenous knowledge for community resilience: Hydro-meteorological disaster risk reduction and climate change adaptation in coastal and small island communities. Jakarta: UNESCO.

IPCC, 2013, Climate change 2013: The physical science Basis. Cambridge: Cambridge University Press.

Kotler, P., & Lee, N., 2008, Social Marketing: Influencing Behaviours for Good. New York: Sage Publications.

Lasica, J.D., 2004, What is Participatory Journalism? [Online]. Available from <http://www.ojr.org/ojr/workplace/1060217106.php> [Accessed on 22 January 2017]

Lester, S., 1999, An Introduction to Phenomenological Research. Toronto: Stan Lester Developments.

Monroe, M.C., Day, B.A. & Grieser, M., 2000, GreenCOM Weaves Four Strands. In B.A. Day and Monroe, M. C. (eds.) Environmental Education and Communication for a Sustainable World. Handbook for International Practitioners., Washington D.C: GreenCOM.

Moser, S.C., 2010, Communicating Climate Change: History, Challenges, Processes and Future Directions. WIREs Climate Change 1(1): pp.31-53.

Mudombi, S., 2013, Role of ICTs in Climate Change Awareness in Seke and Murehwa Districts of Zimbabwe. African Technology Policy Studies Network (ATPS) Working Paper No. 75. ATPS, Nairobi.

O'Neill, S. & Nicholson-Cole, S., 2009, 'Fear won't do it': Promoting Positive Engagement with Climate Change through Visual and Iconic Representations. Science Communication (30) 355-379.

Okenwa, S.N., 2002, The Mass Media: Theories and Realities. Enugu: Bismarck Publication.

Okunna, S.C., 1999, Introduction to Mass Communication, second edition. Enugu: New Generation Books.

Otto-Banaszak, I., Matczak, P., Wesseler, J., & Wechsung, F., 2010, Different Perceptions of Adaptation to Climate Change: A Mental Model Approach Applied to the Expert Interviews. Reg Environ Change. [Online]. Available from <http://edepot.wur.nl/154486> [Accessed on 17 January 2017].

Rose, C., Dade, P., Gallie, N., & Scott, J., 2005, Climate Change Communications – Dipping a Toe into Public Motivation. Available from <http://www.campaignstrategy.org/ valuesvoters/climatechange communications.pdf> [Accessed on 18 January 2017].

Thaler, R. H. and Sunstein, C. R., 2008, Nudge: Improving Decisions about Health, Wealth and Happiness. Yale: Yale University Press.

United Nations Development Programme (UNDP), 2011, Strengthening National Capacity for Climate Change in Zimbabwe. Harare: UNDP.

United Nations Educational, Scientific and Cultural Organization (UNESCO), 2014, Understanding Community Perceptions about Climate Change in the Pacific. Paris: UNESCO.

Weber, E. U. (2010). What Shapes Perceptions of Climate Change? WIREs Climate Change 1(5), 332-342.

ZimStat, 2012, Census 2012 National Report: Zimbabwe Population Census 2012. Harare: ZimStat, Harare.

Chapter Eight

TRADITIONAL KNOWLEDGE AND COMMUNITY-BASED RESEARCH IN CLIMATE CHANGE ADAPTATION IN EAST HIMALAYAS

Lun Yin, *Yunnan Academy of Social Sciences, China*

Misiani M. Zachary, *Nanjing University of Information Science and Technology, China*

Abstract

Agro-pastoralism is important to the Tibetan people from the Eastern Himalayas as a source of livelihood, biodiversity resource management, and culture. In recent years, the impact of climate change has become more obvious, threatening livelihoods and sustainable development. Considering the importance of agro-pastoralism in this context, this chapter presents community-based research on how climate change is impacting Tibetan people. It discusses how traditional knowledge associated with biodiversity is not only an effective toolbox but also a process of adaptation to climate change at a local level. Investigation and action analysis look at seven communities within the Hongpo village from 2005-2010 and the impact of climate change on traditional local livelihood and local response and how their adaptation to it is based on traditional knowledge. A sample of two communities was chosen by virtue of location (altitude, upstream or downstream of Hongpo River). In Nongding, located upstream of Hongpo River and Nongpo, located downstream, 25 households were sampled. At the same time, twenty personal half-structured interviews and five group discussion in these two villages were carried out. The interviews and group discussions involved 25 male older villagers and 30 female villagers. The results reveal that Tibetan people use their traditional knowledge about the climate to guide the sustainability of livelihood development.

Key words: Adaption • Climate Change • Culture • Environmental Change • Tibet

Introduction

It is obvious how climate change will impact on agricultural, water resources and ecological environment (Hardy, 2003). How indigenous peoples or ethnic minorities in polar, mountains, desert, tropical forests, islands, temperate regions observe, understand and adapt climate change is explained, in relation to the reference value and influence of indigenous peoples' traditional knowledge on government climate policy-making (Byg and Salick, 2007). Climate change influences the health sector, heritance, education, traditional knowledge, and the lifestyle of a community. It also influences the strategies put forward by governments, indigenous communities, inter-disciplinary research institute and non-government organizations to adapt to climate change (Green *et al.*, 2009). Climate change has influenced local Tibetans' life and production, local perspectives, and responses to this global phenomenon (Byg and Salick, 2009). Local perspectives on climate change are helpful for the record and protection of traditional knowledge and lay the foundations for the response to climate change and prevention of climate disasters.

Background

The North-West of Yunnan province is a hot spot of bio-cultural diversity. The life of local ethnic minorities is closely related to the natural environment, and the traditional lifestyle depends on natural resources. Notably, the ecological system is extremely vulnerable to climate change, therefore, the impact of climate change on life and production is the most direct and sensitive and local awareness is also more diverse. Hongpo village is located at Deqin County of Diqing Prefecture, North-West of Yunnan Province (98°3'56"-99°32'20", 27°33'44"-29°15'2"N) (Fig. 18). In Tibet, *Hongpo* is a valley that produces silver. Hongpo River is one of the branches of Langcang River (Mekong River). Hongpo village has 7 villages. It has 525 households and 2,379 people, which are all Tibetan. Like other Tibetan tradition, agro-pastoralist is an important traditional livelihood in Hongpo village. Tibetan people have gradually changed their nomadic tradition due to their changing environment as a result of land cultivation by *Han* people and climate change. They've adapted to settled life and also developed agriculture, but they still keep the herding tradition. Since agriculture and herding are both important to them, they finally developed an agro-pastoralist system for their livelihood.

Agro-pastoralist, agriculture, and herding are mutually dependent on each other because they are the main source of livelihoods. They provide basic and necessary food to local people. Herding livestock provides necessary manure for crops, while crop straw provides important fodder to livestock. Additionally, after harvest farmland is transformed to grazing during winter. Tibetans mainly raise yak, cattle yak, yellow cattle, and goats. Animal rearing is dictated by different seasons and location. These can be classified into three types: summer grazing (summer alpine pasture), spring-autumn interim grazing and winter grazing. In Tibetan language, summer grazing land is called *Ru La*, which means grassland covered by snow at an altitude of 4,000m asl and above (alpine meadow).

Figure 8.1: Location of Hongpo Village

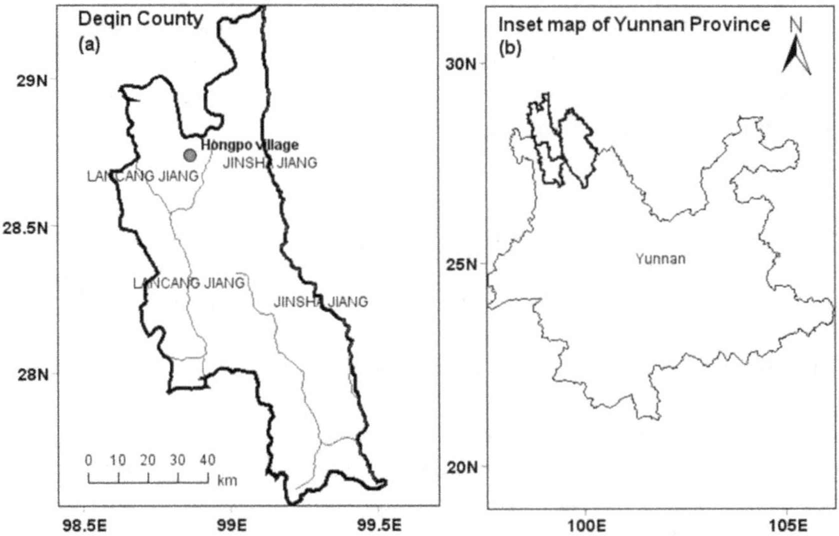

Source: Yin and Zachary, 2017

Spring-autumn interim grazing land is known as *Ru Mei*, which means grassland located at middle altitude. It has an altitude of around 3,000m asl (meadow and sloping land). Finally, winter grazing land is *Ru Bo*, which means grassland located nearby villagers' houses. Its altitude is around 2,000m asl (sloping land nearby the village). While villagers herd yak at alpine pasture and goats at winter grazing land respectively, they will herd cattle yak and yellow cattle regularly among three grazing zones. Every year during March and April villagers will change the grazing land for yak and yellow cattle from winter grazing land to spring interim grazing land. From May to July, they will herd them at summer alpine pasture. During August they will herd at autumn interim grazing land, and from September to February, they will herd yak and yellow cattle at winter grazing land.

In Hongpo, main crops grown are highland barley, wheat, corn, potato, buckwheat and creeping wood. Farmlands are divided into *Qing Wan* or *Ri Xing*. *Qing Wan* refers to the farmland that can directly draw water from the Hongpo River. This kind of farmland is gently undulating and can be planted by crops two times one year. Villagers cultivate the farmland intensively. At the same time, *Ri Xing* or *Ni La* refers to sloping land without irrigation. This kind farmland locates at high altitude and is steep. It can be planted only once a year. *Ri Xing* is originated from slash-and-burn cultivation and still keeps such characteristics as depending on climate, crop rotation, and so on. Usually, villagers plant a large area of crops, but can only harvest a small amout. Every year in May (April, lunar calendar) villagers will plant corn in *Qing Wan*, and in October (November, lunar calendar) they will plant highland barley and wheat. In *Ri Xing*, they will plant potatoes, some buckwheat, and creeping wood. At the same time, they will also plant some vegetables, beans, and chillies near to their houses.

Methodology

Investigation and action analysis on the impact of climate change on traditional local livelihood and local response and adaptation based on indigenous knowledge were carried out from 2005-2010 in 7 communities that cover Hongpo village. Only two communities were selected for this due to different locations (altitude, upstream or downstream of Hongpo River). Nongding, located upstream of Hongpo River and 3,100m asl has 37 households. The other community is Nongpo, located downstream and 2,400m asl, with 29 households. In both of the communities, 25 households were sampled. At the same time, 20 personal half-structured interviews and 5 focus group discussions were administered in these two communities. All of the interviews and group discussions involved 25 males and 25 female older villagers. It reflects a different recognition and response to climate change through quantitative and qualitative analysis (field notes). From the interviews and group discussions, quantitative data was generated on the number of farmers, crops cultivated and livestock raised in both communities. The interviewees were chosen randomly. In order to help Tibetan people adapt to climate change, based on indigenous knowledge and protect natural resources and biodiversity, there was cooperation with villagers, local government and other stakeholders to establish community organizations, training, and community-based investigation. It also worked to promote local people's recognition of climate change and strengthen their capacity to adapt to climate change and biodiversity conservation.

One of the above activities was to establish the Climate Field School (CFS), which was formed by local experts and normal villagers with traditional knowledge. Thus, interviews and group discussions were part of the CFS. Starting from 2006, the main work of the CFS was to find out traditional knowledge related to climate change and collect traditional crop seeds and seeds of Tibetan medical herbs. Through fieldwork carried out by indigenes themselves, CFS disseminated traditional knowledge and seeds among villagers. At the same time, CFS also organized climate change training workshop participated by villagers, local government officials, monks, and other stakeholders (Fig. 9.2). Through the training workshop, local government officials understood the challenges to local livelihood brought about by climate change, and monks understood the relationship between climate change and local traditions and culture. Through statistics, 199 villagers (including 71 females) from 7 communities in Hongpo village participated in fieldwork and the training workshop.

Figure 8.2: Climate Change Training Workshop

Source: Lurong Sinan, 2010

After finishing the above investigation and training, the local government agreed that they would allow CFS to have use-right of 20 mu (a unit of area, 1 mu = 0.0667 hectares) of mountain land as an experiment base for traditional livelihood innovation. At the same time, the local monastery also allowed CFS to carry out protective plantation on the Sacred Mountain. A greenhouse shed was provided to 26 Tibetan households for livelihood innovation. Based on stakeholders' support, CFS carried out the practice of adaptation to climate change in Hongpo village by local people, which has two aims as follows.

The first aim is to mitigate the degree of climate change and its degree of damage at Hongpo village level. The main strategy is to safeguard local climate conditions based on traditional knowledge and prevention from disasters caused by climate change. Safeguarding local climate conditions in Hongpo should focus on the recovery of local vegetation. In the past, mountains in Hongpo were densely covered by lush vegetation. However, from the 1970s to 1990s logging timber was the main industry for local economic development which dramatically damaged large areas of mountain forests, rendering whole landscapes barren. Fortunately, local people have the tradition of protecting Sacred Mountains near to temples. In order to recover local vegetation CFS interviewed many elder villagers and local experts to collect and breed traditional saplings. It also carried out an investigation into the Sacred Mountains by collected tree species and saplings from different altitudes. Based on above two activities, CFS chose 22 saplings as targets for protective recovery. In order to protect these saplings, CFS also invited monks to carry out ceremony of sealing a mountain and built a white pagoda. According to local religion and belief, after sealing, the mountain will be treated as Sacred Mountain, and activities such as collection, cutting trees and hunting are forbidden. Indigenes have also traditionally planted walnut trees near to the village. In order to protect some old walnut trees for over hundred years, CFS also promoted the planting of walnut trees around farmlands. At the same time, CFS helped villagers to make local regulations to manage the trees and prevent them from being eaten by domestic animals.

Natural disasters caused by climate change in Hongpo village are mud-rock flow, flooding, and drought. In recent years, global warming has caused more natural disasters during the raining season. Recovery of vegetation is a very important measure in order to prevent disasters. Additionally, CFS investigated Hongpo River carefully and confirmed that several parts of the river embankments are highly vulnerable (can collapse). The CFS then helped villagers strengthen these areas by using huge stones and cement, which effectively prevented farmland from flooding. During the drought season, CFS helped villagers build water tanks and water channel for irrigation, based on local regulations on water use and management. As traditional water channels were built on earth that easily collapses, CFS helped bury pipes underground to draw water with good results.

The second aim is to adjust traditional livelihood suitably to adapt to climate change. The main strategy is to change and use traditional knowledge innovatively. Besides making some changes to agricultural and herding calendars (lunar calendars) and choosing species of crops and livestock, CFS also helped carry out innovative practice on traditional knowledge. Local experts from CFS collected and researched Tibetan medical herbs, according to different seasons, altitude and shady or sunny slope of mountains. They collected Tibetan medical herbs during spring, summer, autumn and winter seasons respectively. During spring season local experts carried out research and collection in river valley area with altitude under 2,500m asl. During the summer season, they focused on forest area with an altitude between 2,500-3,500m asl. During the autumn season, they moved to areas of alpine meadow with altitude from 3,500-4500m asl. Through a period of one year, local experts collected 146 Tibetan medical herbs, 48 from low altitudes, 57 from middle altitudes and 41 from high altitudes (Fig. 8.3). At the same time, they tried to plant 23 Tibetan medical herbs on the 20 mu mountain with the approval of the local government (Fig. 8.4).

Additionally, CFS supported local experts to establish veterinarian clinic, and invited professional veterinarians to give training to local experts on how to treat and heal new kind of livestock diseases, and planting herbs for a veterinarian. At the same time, CFS supported female villagers in planting some fruits, vegetables, melons and hot pepper in greenhouses to develop the garden economy. These fruits and vegetables are originally grown at low altitudes and warm climate. However, in greenhouses they can be planted at high altitude.

Fig. 8.3: Sample Plant of Tibetan Medical Herb

Source: Ciren Sangzhu, 2010

Fig. 8.4: Plantation of Tibetan Medical Herb

Source: Ciren Sangzhu, 2010

In the past, grapes could only be planted at a low altitude of Hongpo village. From 2003 on, some villages locating at middle and high altitude also tried to plant grapes. Till today every household in these villages plants about 2 mu of grapes. CFS provided relevant training on the grape plantation to villagers. Finally, in some villages which become warmer in winter season because of global warming, villagers want to develop tourism for livelihood. CFS also provided training on eco-tourism and tourist guide.

Results and Discussions

The impact of climate change on local people's traditional livelihood is clear. The impact can be analyzed from the positive and negative aspects. To some extent, global warming has some positive impacts on some cash crops in cold areas. Traditionally, indigenes could not plant those cash crops that needed milder climates, such as grape, apple, hot pepper and so on. Affected by global warming, however, now they can plant these cash crops to increase their income. Global warming can shorten production period of cash crop trees and save time. In these two communities, some cash crop trees' germination and ripening period have been shortened to 15 days compared to 20 years ago. At the same time, because of global warming, it alleviates negative effect from cold weather. Indigenes can also develop eco-tourism to increase their income. Comparatively, negative impacts outweigh positive impacts on traditional agro-pastoralists.

Species Change

Yunnan Province is rich in biodiversity. Its altitude gap is very huge. There are many rivers and cross mountains. A vertical gap of climate change is obvious. Currently, this area is the most protected well zone of biodiversity at the same latitude in the world. It is named as natural gene pool. Hongpo village is also rich in species of agricultural crops and domestic animals. Field investigations revealed that there are 10 main species of crops, among which are sub-species of highland barley. The highland barley is frost resistant. Five sub-species of highland barley are found in Hongpo. In the two villages, through investigation, we find that species of some crops are being affected and changed by global warming. Although we do not have data from before 2005, by comparing data from 2005 with data from 2010 we can see a change in the number of farmers who planted different crops in these two years. The number of farmers who planted frost resistant highland barley in 2010 was much less than those in 2005. At the same time, however, the number of farmers who planted other crops that are not cold-resistant in 2010 increased. Altitude comparison revealed the number of farmers who planted frost resistant highland barley in Nongpo locating at low altitude was much less than the number of farmers from Nongding located at high altitude. In Nongpo, some inhabitants already stopped planting some sub-species of highland barley. However, the number of farmers who planted other crops that are not frost resistant in Nongpo were more than those from Nongding.

In animal husbandry, there are 5 main species of livestock reared by local people (Fig. 8.5, 8.6), among which yak and cattle yak are cold-resistant. Through investigation in the two villages, we see livestock species are also changing. By comparing data from 2005 with data from 2010, it shows the number of livestock reared by every household increased and there was a change in the tendency of species. Firstly, through comparison we see that the number of yak and cattle yak decreased in every household in 2010 compared to 2005, but numbers of yellow cattle increased instead.

Fig. 8.5: Number of farmers from Nongding

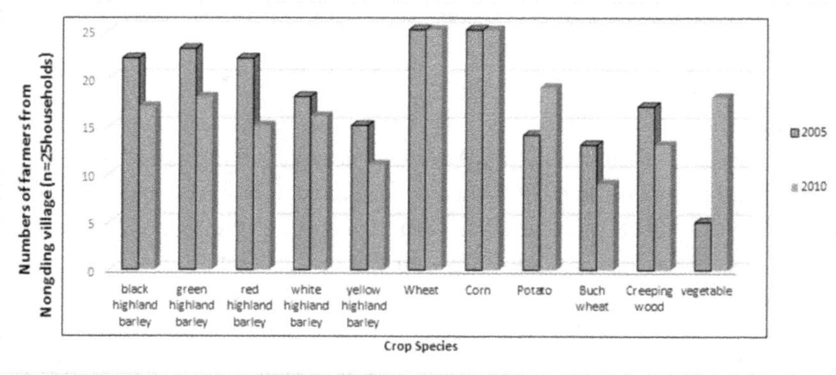

Field data (2005 and 2010)

Fig. 8.6: Number of farmers from Nongpo

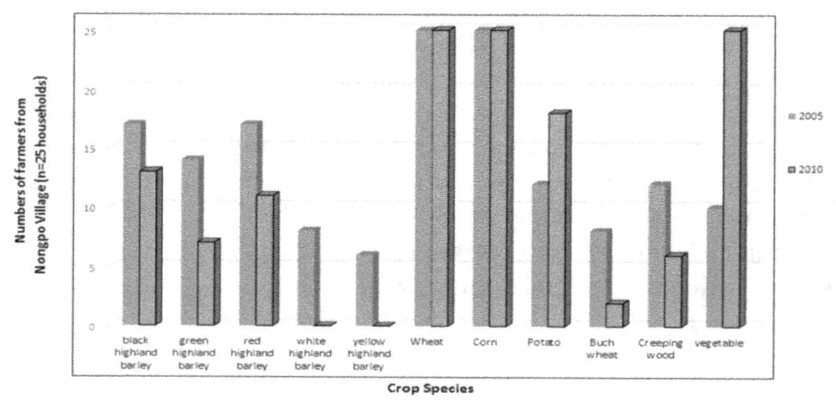

Field data (2005 and 2010)

At the same time, the number of pigs and chicken did not change. Secondly, through altitude comparison, we see that numbers of yak and cattle yak reared by every household from Nongpo Village locating at low altitude decreased compared to numbers from Nongding Village locating at high altitude, but numbers of yellow cattle, pigs and chicken increased instead (Fig. 8.7, 8.8). Some villagers from Nongpo Village already stopped raising yak anymore.

Fig. 8.7: Livestock raised by households from Nongding

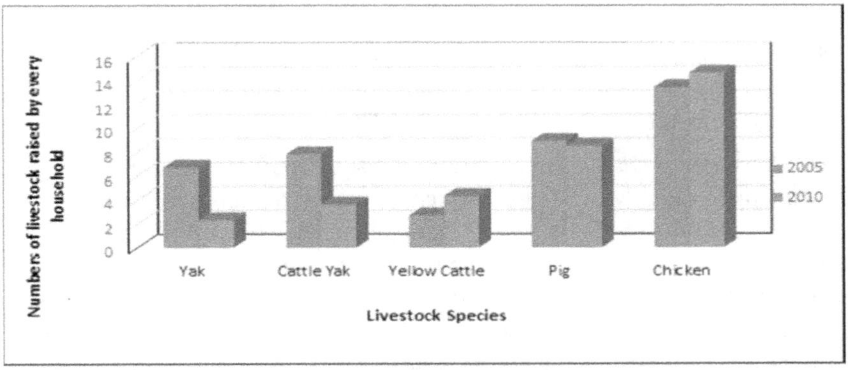

Field data (2005 and 2010)

Fig. 8.8: Livestock raised by households from Nongpo

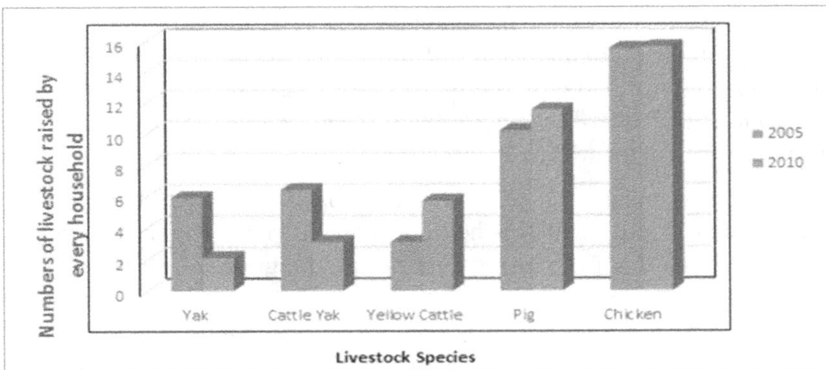

Field data (2005 and 2010)

Environmental Change

The impact of climate change on agriculture and local people's livelihood and the environment is huge. Firstly, because of climate change and global warming, there have been changes to crops' growing period. At the same time, because there is less rain or even no rain during the rainy season, it also negatively affects crop growth. In Hongpo, because of global warming, villagers have to sow seeds in advance. Since the crops' growth period is shortened and growing speed is accelerated, indigenes could not follow their tradition to plant and harvest crops like before. Field observations in the past 10 years reveal that the time for villagers to sow seeds and harvest has been advanced.

For instance, time to sow seeds and harvest corn and wheat has already been shortened by twenty days.

Secondly, climate change brings damages on villagers' farmland infrastructure. Flooding, drought, snow disaster and mud-rock flowing always damage villagers' farmland and their irrigation installations.

Thirdly, climate change causes crop diseases. There are two crop diseases happening frequently in Hongpo, which are wheat rust (Tibetan language *Sa Ni*) and awnless wheat disease (Tibetan language *Ge Na*). Due to global warming, the two crop diseases happen more and more frequently and its degree of damage becomes more and more serious. A large area of farmland has been affected by these diseases. From 2008 to 2010, 7 communities in Hongpo were affected by the two crop diseases, and villagers failed to harvest from their farmland.

Fourthly, climate change also causes plant diseases and insect pests. Global warming creates suitable conditions for insect pests. Since there are less rain and snow than before, it prevents some insect pests from dying in the winter season. In recent years, there has one new pest affecting farmland called *Si Me*. Comparing with traditional pest, *Si Me* damages crops, especially wheat much more serious than other pests. This pest hides underground until December (lunar calendar November), then it will eat wheat root that directly causes the death of the wheat.

The impact of climate change on animal husbandry and local people's livelihoods is also glaring. Firstly, the herding period has been shortened. Migrating period of herding from spring-autumn interim grazing land to summer grazing land has been shortened to 10 days than before, in terms of advancing from middle to end of March. At the same time, migrating period from summer grazing land to winter grazing land has also been shortened, in terms of advancing from beginning of September to the latter of September.

Secondly, growing period of forage grass has also been shortened, because of global warming. Inhabitants from both communities found that forage withers much easier than 10 years ago during winter. Less forage, in turn, affects animal husbandry. The livestock cannot grow as fast as before. Milk products have also decreased.

Thirdly, livestock diseases have become more frequent. Because of global warming, livestock diseases caused by parasite and schistosomiasis happen more than in recent years. Usually, these two diseases spread in summer, but now they also spread in winter. Foot and mouth disease (*Aphtae epizooticae*) is spreading rampantly as a result of global warming.

Impact on Indigenous Knowledge of Agro-pastoralists

As their traditional livelihood, local people accumulate rich indigenous knowledge of agro-pastoralism. It reflects their recognition of natural environment and utilization of natural resources. Along with climate change, however, their knowledge is facing challenges and change. There are two aspects reflecting changes of livelihood knowledge in agriculture caused by climate change.

Firstly, the law of solar term has been disturbed by global warming. This, in turn, affects the periodicity of lunar calendar that is important local knowledge explored and accumulated by villagers based on their recognition on the solar term. Accordingly crop production and harvest have decreased. One inhabitant-*Nu Rongsi* told us an old adage: *'the height of wheat in January is like the height of one chicken'*. In recent years, however, the height of wheat in January is much higher than before that it can cover two chickens' height. It reflects that the growing period of wheat is changing. Additionally, indigenes found that farmland is attacked more easily by pests and it is usually the land that does not follow the lunar calendar and must be planted in advance.

Secondly, since villagers cannot practice local knowledge, crop productivity and quality has greatly decreased during the last ten years. For instance, today's wheat and corn are not as nutritious as before. Villagers think that it is because of raining. Rainwater is very important for crop growing. Different season's raining has different effects on crops. According to local knowledge, rainwater in June has rich nutrients that can accelerate crop growing, while rainwater in August can promote crop maturity. Villagers call the rainwater in August mature rainwater. Villagers believe that during these two periods (June and August) crops heavily depend on natural rainwater and irrigation water drawn from the river cannot replace natural raining. Rainwater and irrigation water from the river have different effects on crops. In recent years, however, because rainwater becomes less and sometimes there does not have any raining, villagers have to draw water from the river to replace rainwater, which directly decreases crop productivity and quality.

Change of livelihood knowledge in animal husbandry is well shown by the change of herding period. Today indigenes cannot follow their tradition to herd livestock based on herding calendar. It takes a shorter time for forage grass to become mature and whither in advance. As a result, villagers have to herd livestock in advance, in terms of moving livestock up or down to the grazing land earlier. Villagers have to herd yak and cattle yak that like cold weather from winter grazing land to summer grazing land, in order to avoid livestock diseases. The close relationship between agriculture and animal husbandry, change of agricultural calendar will affect the change of herding calendar. For instance, usually after finishing harvest of corn, the corn land is used as part of autumn and winter grazing land for villagers to herd livestock. But now harvest of corn is ahead of time, so villagers have to herd livestock to leave summer grazing land for autumn or winter grazing land.

Cultural Effect

Agro-pastoralism is not the only livelihood for Tibetans. Their tradition and culture are also important. Climate change causes changes in species, environment, and indigenous knowledge. The culture of agro-pastoralism is also affected and challenged by these changes. The symbol of ethnical identity of Tibetan people is greatly challenged. In Tibetan culture, highland barley and yak are not the only crop and livestock planted and reared. To some elders, there will be no Tibet if highland barley is not grown and yak are not reared. Tibetan customs and traditions are changing. In agriculture, change in wheat periodicity will affect the agricultural calendar. For instance, during the past villagers would organize a match of shooting arrows during spring festival on wheat farmlands when wheat germinates. This match activity has already been cancelled because wheat germination period is shortened. When villagers plan to organize the match on the wheat farmlands today, wheat has already grown from the beginning period of germination to the third or fourth period of germination because seedlings will be damaged directly.

Food and drink culture is changing. Tibetans use highland barley to make *Zanba* (a staple food), and yak milk to make butter because they believe these two foods can make people strong and healthy. In recent years, however, wheat and rice are gradually replacing *Zanba* and yak butter as local foods. Tibetans belief and religion are also affected. On many religious festivals and ceremonies of Tibetan Buddhism, villagers will throw highland barley and *Zanba* and pray for peace. It is really difficult to imagine that Tibetans will use rice or wheat to replace highland barley and *Zanba* in their religious ceremonies. Unfortunately, this phenomenon is already happening in few households.

Conclusion

Agro-pastoralism is not only a simple way of livelihood for East Himalayas Tibetan communities. It also a method to use natural resources and a comprehensive reflection of traditional knowledge and culture. Agro-pastoralism maintains local people's existence and biodiversity. The most important value of it is to disseminate Tibetan culture. Species, environment and traditional knowledge of agro-pastoralists are affected. Climate change can be mitigated at a local through local people's efforts and adaptation based on their traditional knowledge; secondly, innovation and change of some traditional knowledge can help local people adapt to climate change to some extent. Traditional knowledge and culture of Tibetan people in adapting climate change also provides very important information for the government to make policies for climate change in the future. Traditional knowledge not only provides new space to understand the impact of climate change on our society and environment but also helps us analyse threats, challenges, and changes caused by climate change. It will help local people adapt to climate change. Accordingly, the community-based research project will carry the relevant activities in the future based on Convention on Biological Diversity, in order to strengthen the understanding of the relationship between climate change, traditional knowledge, and biodiversity.

Acknowledgement

Special thanks must be given to Asia-Pacific Inter-Government Research Network, UNDP and UK Ambassador to Chongqing for the support.

References

Byg, A. & Salick, J., 2007, Indigenous Peoples and Climate Change, A Tyndall Centre Publication, Tyndall Centre for Climate Change Research, Oxford.

Byg, A. & Salick, J., 2009, Local Perspectives on a global phenomenon- Climate change in Eastern Tibetan villages, Global Environmental Change (19) 156-166.

Green, D., Jackson, S. & Morrison, J., 2009, Risk for Climate Change to Indigenous Communities in the Tropical North of Australia, Commonwealth of Australia.

Hardy, J.T., 2003, Climate Change-Causes, Effects and Solutions. John Wiley & Sons Ltd, London.

Chapter Nine

CONCLUSION

Vincent Itai Tanyanyiwa, *Zimbabwe Open University, Zimbabwe*

Suiven John Paul Tume, *Green Care Association, Cameroon*

Climate change is a global as well as a national issue which affects the biological, physical, social and economic sector of the environment. Researchers have applied different methods for assessing climate change perception. They use questionnaire survey, field observation, key informant interviews, focus group discussions, climatic data collection and modelling for analysis of climate change perception. People have diverse perceptions about seasonal changes. Climate change has caused a decrease in agricultural production, reduced sources of water, adverse health impacts and changed the livelihood patterns to name but a few. To reduce the impact of climate change, there is a need for accurate climatic data collection, analysis, communication, forecasting, effective early warning system, vulnerability assessment and awareness. This book presents first-hand experiences of the concept of climate change perception in Africa and South Asia. It draws case studies from Cameroon, Kenya, Zimbabwe, Nepal and the East Himalayas. The Cameroonian experience is based on stakeholders in climate change communication. In Kenya, the experience of climate change communication focuses on practical field experiences based on climate change hearings. The cases from Zimbabwe are based on the role of social workers in climate change mitigation, the use of weather information for decision-making and game changers in climate communication. The role indigenous knowledge and community-based research in climate change adaptation in East Himalayas as well as the perception of climate change in Nepal present first-hand research from South Asia. Indigenous knowledge is very important because it is grounded in local traditions and therefore it is fairly well understood by the locals.

The role of Social Work in rural development and the climate change mitigation agenda is critical. African Social work embraces the curative or remedial approach to solving social problems during its inception. Majority of NGOs employed social workers are involved in developmental, social work, with a specific bias to rural areas and poverty having the responsibility to mobilise communities to take an active interest in the problems affecting them. The International Federation of Social Workers (IFSW), the International Association of Schools of Social Work (IASSW) and the International Council on Social Welfare (ICSW) note that people health and wellbeing suffer as a result of inequalities and unsustainable environments related to climate change, pollutants, war, natural disasters and violence to which there are inadequate international responses. Social work institutions have scholars and have been producing professionals who lack the capacity to address the nexus of environmental justice, social work, and social justice. This is an area that needs further research especially by developing countries.

Most rural societies are relatively integrated. Various components of life-agricultural and non-agricultural, economic, social and political, religious and secular closely interrelated and interwoven. Rural people do not easily recognize planners, extension workers and other government officials' distinctions make between responsibilities of different agencies or economic and social planning because in their own lives all these things affect each other. A study in the Chimanimani District of Zimbabwe analysed interface between academic researchers, development practitioners, district institutional actors and local communities in a project aimed at forest resources management enhancement. It noted that for involvement and accommodation of various actors' interests quantitative and qualitative action research strategies and participatory geographic information systems approach are critical for effective environmental management.

Farmers cannot access reliable and usable weather data. Information is often unavailable. If it does exist, the quality is poor, or it is inaccessible to those who need it most. This is because the information that is given is not user-friendly therefore in most cases not easily accepted by the end users. Farmers do not get reliable information on drought forecasts, rainfall distribution, and pest outbreaks. Most governments, especially in developing countries as well as some development agencies, do not recognize and prioritise the value of climate and weather data. This has stifled investment in infrastructure and proper functioning of state institutions charged with collecting and serving climate and weather data. Zimbabwe alone can tap into the USD$10.2 billion Green Climate Fund adopted in 2011 by the UNFCCC to finance adaptation and mitigation projects and programmes in developing countries. Financing infrastructure on basic meteorological observations to generate weather and climate data fit for agricultural purposes should be a priority for the fund. Though many players are involved in the generation and dissemination of weather and climate information, the Zimbabwe Meteorological Service Department (MSD) remains the main source of seasonal weather forecasts. Since this is a government department under the Ministry of Environment,

Water and Climate. The political and socio-economic as well as the technological challenges have impacted on the generation and dissemination of weather and climate information. There has been a mismatch between farmers' needs and the disseminated information which is usually communicated in English and as such farmers have not been able to effectively use the information in the decision making the process as much as they would want to. There is, therefore, need on the part of MSD, Zimbabwe to change on the way they disseminate weather and climate information.

Local and external actors coalesce despite the social differences usually defining social interfaces as technical expertise, material and financial resources possessed by external actors. External actors, despite their technical and financial resources, lack knowledge about the local area. Actors as local authorities, government departments, quasi-government organisations officials and local communities might lack certain technical, financial and material resources, but have enormous knowledge about their areas, and are passive recipients of the packages that come with external actors. Social workers in rural development and climate change interventions can bridge the gap and harness synergies between various actors for desired outcomes of climate change adaptation and mitigation.

Kenya through its various formal and informal engagements from its academic, civil society, youth, private and public sectors has utilised all opportunities to exchange and share climate change information. The work highlighted in this book is a snapshot of the kind of effort made thus far in doing so to a growing audience whom it is hoped then go on to share the insights they glean with their own ecosystem of constituents and/or communities. Due to the overwhelming body of work that continues to emerge, all the aforementioned avenues will be useful in Kenya's endeavour to have a population that is able to build resilience while integrating low carbon development mechanisms effectively.

To most of Tibetan people living in the mountainous area of the Eastern-Tibetan Himalayas, agro-pastoralism is not only a simple way of livelihood, or a method to use natural resources but is also a comprehensive reflection of indigenous knowledge and culture. Agro-pastoralism maintains local people's existence and biodiversity of crops and domestic animals for inheritance. The most important value of it is to disseminate Tibetan culture which has strengthened climate change adaptation and mitigation efforts, species diversity, environment and indigenous knowledge systems

Community-based research is critical in analyzing the relationship between climate change and indigenous knowledge systems. Community-based research does not only provide new space to understand the impact of climate change on society and the natural environment but also helps to analyse threats, challenges, and changes caused by climate change along ethnic areas. It will help local people adapt to climate change. Accordingly, the community-based research project will carry the relevant activities in the future based on 'Convention on Biological Diversity', in order to strengthen understanding on the relationship between climate change, indigenous knowledge, and biodiversity. Through these initiatives, the project will explore theory, logical structure and methodology to adapt to climate change based on indigenous knowledge, improve stakeholders' understanding on indigenous knowledge, and find innovation and practice on adaptation to climate change. Based on above background, the project will carry out research on innovation and practice on adaption to climate change based on indigenous knowledge, and explore theory and methodology about adaptation to climate change based on indigenous knowledge.

It is difficult to study impacts of climate change due to a lack of long interval instrumental climatic data. Accurate data collection and forecasting, effective early warning system, adequate mapping of hazard and vulnerability, and awareness generation would minimize risks of climate change. Use of water in hydropower generation, irrigation and storage reservoir would help to minimize climatic risk. To address the adverse impacts of climate change, an integrated and coordinated approach to improving understanding of climate change science, impacts, vulnerability, adaptation as well as mitigation should be prioritized. There is a need for awareness, advocacy and negotiations to minimize climate change impacts and increase resilience. For enhancing climate change adaptation options, economic, social and ecological dimensions of sustainable development should take centre stage. Impacts and adaptation strategies towards climate change need to be identified with proper planning processes. Shifting of habit towards low carbon economy could minimize adverse impacts of climate change. Indigenous knowledge needs to be documented, verified and replicated in different regions of the world especially in developing countries where IKS is more prevalent as compared to the developed north.

Climate change communication is gaining ground in developing countries. All the media houses surveyed in Northwest Cameroon were fully aware of the vulnerabilities and direct impacts of climate change around their local areas. There is a need for more interactive radio programs based on climate change and the environment especially for the media houses that do not air such programs. Journalists need more capacity building and training on climate change and environmental reporting. Journalism clubs should be encouraged in schools, colleges, and universities where students can disseminate environmental issues to their peers. Radio quizzes should be aired during all climate and environment-related international days. Socio-cultural, religious gatherings like churches should be used as communication channels on the dangers of unsustainable activities such as UN environmental calendar, sketches and poems should be used as a means to get across climate change messages. Hands-on-the ground projects like tree planting should be encouraged in communities. NGOs and schools can help in this endeavour. Traditional address systems should be used to sensitize the rural populations on the dangers of their activities like overgrazing, slash and burn farming and bushfires. Newspapers have a duty to dedicate columns for climate change and the environment. Use of local languages to reiterate causes of climate change, vulnerability and adaptation should frequently be aired through radio stations. Posters in public spaces on the dangers of climate change should take precedence. Sensitisation and workshops at grassroots should be encouraged. The youth should be fully engaged in climate change communication through popular social media and community forums.

Climate change communication targets specific audiences who, in most cases, are diverse and heterogeneous. That being the case, there is a need for communication on climate change to be packaged in such a way that it is accepted by the targeted audience who will, in turn, it is anticipated, will eventually change their behaviour. That being the case, messages have to be packaged in such a way that the intended objectives are achieved, which will contribute to adaptation and mitigation to climate change. There are various factors which hinder the success of climate change communication. These factors include the targeted audience either not having access to the information, or when they get access to the information, it tends to be too technical and beyond their comprehension. Overall, there is a need for information on climate change to be packaged systematically taking into consideration the needs of the target audience. Furthermore, the wide contemporary communication platforms such as social media are not effectively utilised for climate change communication. This platform contributes to the ineffectiveness of the climate change communication in the country. Instead of effectively using such platforms to disseminate important information, communication on social media is never taken seriously in African countries; rather it is mostly used for entertainment and amusement. However, a change in attitude may go a long way in changing this current scenario of looking at social media as second fiddle to mainstream media.

www.ingramcontent.com/pod-product-compliance
Lightning Source LLC
LaVergne TN
LVHW010721210926
848485LV00007B/1404/J
* 9 7 8 1 6 2 2 7 3 3 0 8 8 *